SpringerBriefs in Physics

SpringerBriefs in Physics are a series of slim high-quality publications encompassing the entire spectrum of physics. Manuscripts for SpringerBriefs in Physics will be evaluated by Springer and by members of the Editorial Board. Proposals and other communication should be sent to your Publishing Editors at Springer.

Featuring compact volumes of 50 to 125 pages (approximately 20,000–45,000 words), Briefs are shorter than a conventional book but longer than a journal article. Thus, Briefs serve as timely, concise tools for students, researchers, and professionals.

Typical texts for publication might include:

- A snapshot review of the current state of a hot or emerging field
- A concise introduction to core concepts that students must understand in order to make independent contributions
- An extended research report giving more details and discussion than is possible in a conventional journal article
- A manual describing underlying principles and best practices for an experimental technique
- An essay exploring new ideas within physics, related philosophical issues, or broader topics such as science and society

Briefs allow authors to present their ideas and readers to absorb them with minimal time investment. Briefs will be published as part of Springer's eBook collection, with millions of users worldwide. In addition, they will be available, just like other books, for individual print and electronic purchase. Briefs are characterized by fast, global electronic dissemination, straightforward publishing agreements, easy-to-use manuscript preparation and formatting guidelines, and expedited production schedules. We aim for publication 8–12 weeks after acceptance.

Tamás Sándor Biró

Variational Principles in Physics

From Classical to Quantum Realm

 Springer

Tamás Sándor Biró
Nano-Plasmonic Laser Inertial Fusion
Experiment (NAPLIFE)
Wigner Research Centre for Physics
Budapest, Hungary

ISSN 2191-5423   ISSN 2191-5431 (electronic)
SpringerBriefs in Physics
ISBN 978-3-031-27875-4   ISBN 978-3-031-27876-1 (eBook)
https://doi.org/10.1007/978-3-031-27876-1

This Springer imprint is published by the registered company Springer Nature Switzerland AG
The registered company address is: Gewerbestrasse 11, 6330 Cham, Switzerland

Man if ever learns it,
will do it in his kitchen.

Imre Madách: The Tragedy of Man

Preface

This book has been written based on a series of lectures given for physics students at the Justus Liebig University (JLU) in Giessen, Germany first, and later to physics and engineering students at the Technical and Economical University (BME) in Budapest, Hungary. Deviating from other traditional courses in theoretical physics curriculum, this course is not organized around a narrow circle of problems or disciplines. To the contrary, here we follow a given principle in thinking using the related mathematical methods, the variational calculus across the classical mechanics, optics, electrodynamics, thermodynamics, and quantum mechanics, both in interpretation and derivation of the known basic equations. This method is versatile, not only in classical mechanics and particle physics courses.

Crossing therefore the usual theoretical physics disciplines we use the variational principle here as a guideline. Not the everyday calculational help is important here, but the role in theory making that variational and symmetry principles play. This appears in the treatise of such details as classical electrodynamics or the Schrödinger equation in quantum mechanics, where most textbooks and lectures ignore the variational principle and start away with the discussion of the field equations.

Meanwhile a derivation of the—in high schools also mentioned—Newton equations of motion from the variational principle of Lagrange function and the related action is part of the university theoretical physics curriculum worldwide. At several places however the discussion of technical details of the Maxwell equations is far earlier than a mention of the Lagrange function of electrodynamics, if that is mentioned at all in hasty courses. The Schrödinger equation occurs usually as a *deus ex machina*, mentioning that its consequences are justified by experiments and modern technology. Students have to learn its use and solution in practical mini problems. Since understanding it in the sense of re-discovery with all the connected emotions and cognitive rewards looks hopeless, its "derivation" or a review of alternative but equivalent presentations became a prohibited heresy. Meanwhile it is a fact that quantum mechanics is the closest compromise to keep as much from classical mechanics as possible. Schrödinger's variational principle formulates exactly a minimization of the deviance from classical Hamilton-Jacobi equation, its non-fulfillment integrated over space and time is the variational action.[1] Schrödinger in his original article presents this variational principle.

Omitting this in several theoretical physics courses might be due to the later dominance of the orthodox Copenhagen school: they felt more important to keep the Hamiltonian exactly describing the energy and they rather use operators instead of functions, than admitting that the old dispersion relation between energy and momenta are not fulfilled in the quantum world. Or alternatively, because the user's attitude has won when facing the mathematical apparatus of quantum physics, the infamous "shut up and calculate" behavior developed. It assumes that we do not have to understand what we are doing, it is enough if it is functioning. Here the author of this book thinks that following the logic of the original variational principle which emphasizes that quantum mechanics is a minimal break with its classical ancestor can be a useful experience for most physicists or students studying physics.

The usefulness of this can be divided into several layers: on one hand it is important for a practicing professional in the higher education to feel that the fundamental formulas of our science are embedded into the human thinking in multiple ways, they can be viewed from different corners and directions and still mean the same. On the other hand, as long as the measurable consequences are the same, these ways are equivalent, equal in rank, and one cannot distinguish one view against the other. And if they would differ in some particular result on a measurable consequence, then experiments can decide which is valid. Knowledge and the interpretation of knowledge is not the same.

The variational principle view of fundamental laws in physics helps to get conscious of global thinking: seeking for an optimum is an ordering principle and as such it unifies various sub-disciplines in a common frame and helps to recognize similarities beyond differences.

Budapest, Hungary   Tamás Sándor Biró
January 2023

---

[1] A more abstract mathematical approach also points out that the classical mechanics can be defined on a distributive network in quantum logic, while quantum mechanics on its closest generalization, on an orthomodular net.

Acknowledgements

First of all T.S.B. owes gratitude towards Dr. László Orosz, docent at the Department of Physics at the Technical University Budapest (BME), for initiating the idea of writing a textbook and for his persistent encouragement, as well as pedagogical advice during the Hungarian edition. To Prof. János Kertész, member of Hungarian Academy of Science (H.A.S.), I express my acknowledgment for his support of my teaching in the framework of the Physics Institute at BME. Experiences from this activity are precipitated in this book.

Professional lector of the Hungarian edition in 2012 was Prof. András Patkós, member of H.A.S., that time head of the Department for Atomic Physics at the Eötvös Roland University (ELTE). His several comments and questions inspired me to increase the precision of some statements, and at a few locations to extend explanations beyond my original plans.

A special thank belongs to József Pálinkás, that time president of Hungarian Academy of Science, and to the Physics Class of that Academy for a financial support for the Hungarian publication of this book.

Important help arrived from my colleagues at the Institute for Particle and Nuclear Physics (KFKI RMKI), Dr. Péter Ván and Dr. Árpád Lukács, who have read my original manuscript before submission.

For their various contributions in theoretical research, acknowledgement is due for Drs. Antal Jakovác, Péter Ván, Etele Molnár, Rogerio Rosenfeld and for Gábor Purcsel and Károly Ürmössy.

For the English edition Dr. Archana Kumari was of indispensable help in commenting from the young researcher's point of view. She deserves acknowledgement.

Finally, last but not least, every scientific research and university teaching work needs a corresponding background: without the exerted patience and understanding from the side of my family this book could not have been written. I hereby acknowledge the support from my wife, Tünde Biróné Riz, my daughters, Szilvia and Réka Biró, and my step-children Katalin and Gábor Csengery.

Contents

Chapter 1
Introduction

In this book, a brief history of the principles that define how a physicist thinks is presented. After reviewing that, the most important elements of the mathematical formalism of the variational calculus with particular regard to the application of the concept of the functional derivations are summarized. In the second, first technical chapter, we deal with the variational principles of mechanics. It is used to solve static (equilibrium) problems, which appear as a generalization of the principle of virtual work (by Bernoulli) and D'Alambert's principle, which serves as the basis for Hamilton's action principle. Various constraints, treated as side conditions, are taken into account by the Gauss principle, and are illuminated with the help of Lagrange multipliers; in the process providing us with the modern idea of the effective potential. This is followed by the analysis of Maupertuis principle; not only as the story of a mistake and its correction, but also as a foreshadowing of the general geodesic motion. So far this principle, when applied to motion, focuses on the shape of the trajectories instead of their temporal evolution. The chapter on the principles of mechanical variation mentions the general action–angle variables and their use. It concludes with a description of Fermat's Least-Propagation-Time principle.

The third chapter deals with gravity, built up on the treatment of the Mapertuis principle applied to a relativistically moving mass point. Similarly, the relativistic motion of electric point charges follow a geometric variational principle, from which the Lorentz force can be derived. The Newtonian (i.e. weak) gravitational field due to the equivalence of gravitating and inert mass, in the temporal coordinate differentials can be interpreted as a curvature effect. This opens the way to the curvature of total spacetime by minimizing the Einstein-Hilbert action, and from this to derive the Einstein equations.

The fourth chapter describes the classical fundamental laws of electric and magnetic phenomena, derived from corresponding variational principles. The Gauss theory of electrostatics, Ampère's law of magnetostatic mechanics and energy carried by the fields with certain additional conditions are derived. In electrodynamics, the symmetry of the magnetic and electric forces (EM-duality) are formulated by

© The Author(s), under exclusive license to Springer Nature Switzerland AG 2023
T. S. Biró, *Variational Principles in Physics*,
SpringerBriefs in Physics,
https://doi.org/10.1007/978-3-031-27876-1_1

constructing a common variational principle. This step intuitively introduces, with a single modification, the Faraday induction and Maxwell's displacement current. The Lorenz gauge fixing is also formulated in the variational principle so that the wave solution emerges with natural simplicity. The electric—magnetic duality and the use of complex or quaternionic variables round off the picture of electrodynamics.

The fifth chapter is about the Schrödinger equation of quantum mechanics, also known as wave mechanics. This equation can be obtained from the variational principle based on the Hamilton-Jacobi equation when its violation is minimized. If the weight function, which weights the "seriousness" or significance of the deviation from zero, is chosen to be so that the Schrödinger equation itself is linear, then it turns out that the only proper choice is the eikonal function associated to the classical action, in a quadratic or for the complex case in a bilinear expression. This action principle behind the quantum wave function is a seldom quoted, implicit aspect. At the same time, it helps to view certain over-philosophized aspects, such as the quantum uncertainty (which according to its original German name "Unschärfe" would be more accurate to call unsharpness, indefiniteness), from a new perspective. That discussion also may serve as a unique introduction to Feynman's path integral formalism or as an introductory example for that.

Relativistic hydrodynamics is the basis of both cosmology and modern heavy ion physics, it is an important theoretical tool in modern physics research. Interestingly, it can also be derived from a variational principle by generalising the Mapertuis principle to media. The variational principles of thermodynamics, although mathematically more primitive, are philosophically more general in their significance than the others mentioned so far; this may justify the inclusion of them in the present book. Entropy as a quantity to be maximized with general side-conditions expressing the conservation of certain other quantities, leads to different thermodynamic potentials. These are constructed by Legendre transformation, where the quantities acting as Lagrange multipliers are related to temperature and to other intensive variables. The maximal entropy principle, also used in informatics and statistical physics, has been generalised in several different ways since its classical formulation by Clausius. These research areas are omitted from the present book.

Although the technique of the variational method was spreading among physicists only after the triumphant procession of the Newtonian (mechanical) world view, similar ideas and principles originate from much earlier, most from the antique period of human history. At the same time these methods frequently re-appear in the modern and contemporary physics, too. In order to pay attention to this history in this introductory chapter, a few additional concepts like symmetry shall be mentioned, before presenting a brief summary of the mathematical background. With the aim that the esteemed Reader shall strongly feel and understand the importance and deep cultural nature of the basic idea of the variational approach, that points far beyond its utilitarian use in quantitative disciplines, like physics.

It is a global problem to search for an optimum (maximum or minimum) of something, already in its formulation. Most surprising is the lesson during the application of variational principles in physics that—assuming the continuity among neighboring paths—that this *global* principle leads to the same result as the *local* mathematics, describing the motion point by point, in small, causal steps. At the same time further circumstances of the evolution of a physical system, in the form of initial, boundary or auxiliary conditions can be taken into account more naturally, clearly, easy to follow and deeply intuitive in the framework of the mathematical formalism applied by the variational method. The determination of an optimum implicitly assumes the view of alternatives, and as such it promotes the understanding of the path integral concept, and in general the functional integrals summing over alternative orbits.

1.1 Brief History of Roots

The variational method has its ancestors. The Laws of Nature, in particular that of the motion was in the focus of natural philosophy already at the dawn of the European culture, which had roots in the East, in the first civilizations. The fight between calculability and the paradox nature of motion was long part of the evolution of ideas.

Here we review a few stations of this evolution leading to our modern physics and the central use of variational principle in its heart. The sampling is almost random, but a skeleton of the arrow of time should be seen in it.

1.1.1 Antiquity: Paradoxes

Our knowledge dates back to the old Greeks. For long time the European thinking was determined by Aristotle's principle, cited as "horror vacui" (angst from the nothingness), that from the role of highest authority advanced later to the most criticized one, equally unjustified. To date it is no more easy to reconstruct Aristotle's original thoughts. The phrase "horror vacui", the abhorrence from the vacuum is frequently interpreted nowadays as he would have regarded a sucking force of the vacuum as the cause of motion.

Transgressing this vulgarized interpretation, questioned with right, it is worth to consider another, more extended reconstruction. For the men of the antique, and so for Aristotle, the vacuum was the archetype of paradoxical and therefore impossible things: the "nothing", about which we talk as if it were "something" located "somewhere". Can the nothing exist? According to them not, since the question itself is paradoxical. The "horror vacui" principle in this context meant that phenomena described by contradictory arguments cannot be realized in nature, "the contradiction is not a real existing thing". To be free of contradiction is even until today a requirement for scientific theories. According to experience during hundreds of years we even believe that this principle expresses one of the most important qualities of the reality.

A further development of this principle, connected to the Pythagoreans, sounds more like aesthetics than experience: the nature is "beautiful", the universe is harmonic, therefore its description must be based on beautiful and harmonic laws. According to Pythagoras, science must describe everything by rational numbers or in the language of geometry. This strict requirement is already abandoned in physics, but still, the inappropriate numbers are called "irrational", meaning "meaningless" or "disproportionate". Like the square root of two: it existed as a geometrical object, as the diagonal of square with unit length sides, but not as a ratio of two integers.

By all means, the sophists, first of all Zenon, also have to be mentioned here. The essence of their derivations of arguments was to demonstrate the paradox nature of (continuous) motion. Up to our today's knowledge this statement shows a point. The conclusion that what is paradoxical is also impossible, and therefore in reality there is no motion, at most its illusion, is contradicting the experience. We should not forget that the paradox emerges from dividing the motion to smaller and smaller pieces and not caring for the division of time accordingly, as opposed to the view of comparing the initial and final state of a given motion. These views can only be harmonized in the framework of infinitesimal mathematics, pioneered later.

Into that direction already Archimedes made first steps by seeking for formulas of calculating volumes and surfaces of geometrical bodies and by determining centers of weight for variously shaped masses. However, Newton was the first to achieve breakthrough, and Leibniz, Euler and Lagrange made this technique understandable and standardized much later.

1.1.2   Medieval Thinking: Transition

The most known intellectual successor of Aristotle in the medieval age was Thomas of Aquino. He was most interested in the question: Does motion has a goal? In our modern context he might have had asked whether all physical motion can be described within a local theory, using causal chains. This obviously depends on the system under study. In more complex systems embracing several factors it is not rare to observe a memory effect, that kills the locality in time. And we know that the clever rat, who remembers where is the cheese placed in a labyrinth, performs a motion which looks aimed and purposeful.

It occurred as obvious for the medieval thinkers that humans and animals move tendentiously, and dead bodies do not. It became famous an example about the burro of Buridan: the donkey finds himself between two fields, one with wheat the other with barley. It feels hungry and loves both delicacies, it would run to the nearest field. But whenever its position is symmetric, equally far from both fields, the burro cannot decide and the purposeful motion is impossible. The donkey—since it is a donkey—stays motionless and starves. What here emerges is the principle of symmetry connected to the theory of motion and the breaking of symmetry as an attribute of reality. At the same time the paradoxical formulation of the story reminds to the great ancestors in the antique. The summary under the line is nevertheless pointing out the indefensibility of the purposefulness in the description of physical motion.

1.1.3 Renaissance: Dynamics

Renaissance means being born anew; the birth of modern physics coincides with this period. In line signalled by Kepler, Galilei and Newton the perception and concept of motion had been changed entirely. Meanwhile Kepler's stand was being a court astrologist and Newton had alchemy as his hobby. Their basic principles about motion are until to date the fundamental pillars of physics taught in schools: the uniform, inertial motion itself is a natural state without any extra cause, and therefore no absolute motion exists. We have to chase causes only behind the changes in the state of motion, we call these causes "force" since then. The purpose of physics is to find and calculate the most beautiful (state of) motion, not the most beautiful state.

The basic principles of Nature (principia naturalis) are causal: the changes develop from point to point, form cause to consequence. The following phrasing of this stems from Huygens: "*all effect is local*". However, the theory of Newtonian gravity did not yet follow this ideal image, it contained action at a distance. Newton himself referred to this as a manque to be removed from his theory.

1.1.4 Enlightenment: The Global View

The age of enlightenment had enriched physics with new principles, first of all due to the works of Euler, Lagrange, Hamilton and Jacobi, and of course Gauss in which a new form of the Newtonian mechanics, but leading to mathematically equivalent results, was elaborated. Our formulas used nowadays are traced back to that period. Then the first variational principles in physics were born.

The action functional takes a calculation of cost, valid for the totality of the motion, as a basis. We feel in this idea the developing world trade, the more and more public attitude on calculating a maximal profit from the counterbalance of expenditure and income. The notions of momentum, energy and velocity were mathematically defined and generalized. The most determining experience is that holistic principles lead to local, causal equations of motion. The opposite side of this coin is that the interplay of local forces lead to a global harmony. The forces of free market will make everyone happy—and whom not, he is guilty himself. It becomes clear in the modern mechanics that certain conditions, constraints, in particular those arising from symmetry, influence the motions to a great extent.

1.1.5 Romanticism: Grand Unification

The 19th century was the age of great theoretical works, grand unifications. Based on Maxwell's equations the electrical, magnetic and optical phenomena are unified in electrodynamics with a still sensible ancestry from mechanics. At the same time this is the first field theory, an important step towards the view of local forces and natural phenomena as played on a multidimensional stage.

Another important development is the crystallization of thermodynamics. Variational principles with varying conditions are interrelated and can be transformed into each other; in the background the mathematics of Legendre transformation hides. Moreover nature has some laws which are expressed in an inequality instead of an equation: the entropy in closed systems cannot decrease, as a trend. Related to this the final state of universe, the "heat death" is visioned.

An imprescriptible merit of thermodynamics is that it raised the attention of physicists to the atomic world, a playground only for chemists until then. The mechanics as a role model was a midwife at the birth of kinetic gas theory. The trust in the fundamental principles is stronger then ever.

1.1.6 Modern Times: Uncertainty

The 20th century brought a revolution in several issues, so did in physics. The quantum mechanics as "new physics" seemingly broke all known rules. However, interestingly the natural laws in quantum and classical physics, in the micro and macro world, not only differ but they also show similarities. The famous Schrödinger equation was derived by him from such a variational principle that optimized the violation of the classical Hamilton–Jacobi equation. Quantum mechanics is the most classical non classical physics. What we thought to know before is untrue, but we shall be somehow happy, that it is untrue only to a least extent.

Another break with the classical tradition in 20th century physics is the theory of relativity. Partially known since Galilei, inherent in the Maxwell equations, but at the first time by Einstein raised to the rank of a fundamental principle is the special relativity principle. A further generalization of it led to the unified geometric view of space, time and gravity. Whether all laws of nature are geometrical? This would not deny the applicability of algebraic or other, e.g. numerical symbols, but it would be superfluous to cite the causes of changes as genuine forces.

1.1.7 Contemporary Adventure

The history of ideas after the Great World Wars I and II, perhaps due to their proximity to our age, shows a fractal, chaotic picture. Yet we grab a few elements from the evolution of fundamental physical principles, even if arbitrarily.

The quantum version of field theory and the physics of elementary particles brought the reconciliation of the quantum and relativity principles to the agenda. Also the measurement technology reached that level where minuscule quantum particles moving with velocities near to that of the light can be studied. New among the principles is the method of Feynman's path integral, which instead of choosing the "best" orbit democratically sum over all possible orbits "contributing to the reality". True, their contribution is restricted by strict rules, and possible paths interfere like waves. Under certain circumstances at the end some possibilities dominate over the others; and exactly then the physical world occurs as "classical".

It is still under discussion, the quest for the Holy Grail of the physics at the turn of millennia, the quest for the original "ancestor" symmetry. Grand unification theories, the "Theory of Everything", quantum gravity and elementary string theory all show a fermentation in the mathematical arsenal. The experimental control is not possible in all aspects, in some other aspects it looks rather discouraging so far. We would wish that a good fundamental principle is both beautiful and it functions in praxis.

1.2   Variational Calculus Basics

The variational principles, the mathematical background to optimum searches, is based on the notion of *functional*.[1] The description of physical states, including that of the state of motion, means the selection of a single or a few ones from close alternatives. The sought result is a function, but the quality of this result, as a measure of the optimality, can often be expressed by a number.

In the followings we deal with the notions and basic properties of the function and functional, as they are used in the present book. We present examples based on the real numbers, $\mathcal{R}$, but most of the formulas—sometimes after a necessary generalization or restrictions—are valid also for complex numbers.

1.2.1   Function and Functional

Let the solution of a physical problem be a rule connecting a real number to another real number, briefly a function, $f : \mathcal{R} \to \mathcal{R}$. This can be the solution of an equation describing dynamics, giving the spatial coordinates as functions of time, or describing a two-dimensional orbit of a moving mass point in space. The set of these functions are noted here by $\mathcal{F}$, the sought solution functions are elements of this set. A (real) functional orders a (real) number to such a function: $I : \mathcal{F} \to \mathcal{R}$. In the followings we are going to use a distinction in the notation of the number valued functions and functionals: the functions are written using round, the functionals using square brackets. In this way the real number $f(x)$ is the value of the function f at the argument x, while $I[f]$ is the value of the functional I for the selected function f.

---

[1] A functional in the mathematics is a mapping of a V vector space over a base body into the base body. The base body can be R, C or even R^n.

A good example is given by the following functional,

$$I[f] = \int_0^1 dx \left(|f'(x)|^2 - x^2 |f(x)|^2 \right), \tag{1.1}$$

those value is a definite integral over an integrand composed using the function, its derivative and in some cases another function of the independent variable. Another example describes the length of a curve, parameterized by the function $f(x)$ in a two-dimensional space, between the points at $x = a$ and $x = b$:

$$L[f; a, b] = \int_a^b dx \left(1 + f'(x)^2 \right)^{1/2}. \tag{1.2}$$

It is worth to be noted here that while $L[f]$ is a functional of the function f, it is at the same time a function of the endpoint coordinates a and b.

1.2.2 Variation

In the classical physics the possible solutions are overwhelmingly continuous and repeatedly differentiable, short smooth, functions. These functions are many and they are dense in the set $\mathcal{F}$. Therefore it makes sense to consider "close" functions, and to talk about a nearby function, $f + \epsilon \in \mathcal{F}$. A well-applicable definition of the "nearby" property for functionals of type $\int L(f, f', x)dx$ is as follows:

Definition 1.1 Let f and ϵ be smooth functions on the interval $[a, b]$. We call the function $f + \epsilon$ near to f if the function norm of ϵ can be made arbitrarily small, e.g.[2]

$$\|\epsilon\| = \max_{[a,b]} \left(|\epsilon(x)| + |\epsilon'(x)| \right) \longrightarrow 0^+,$$

while in the endpoints there is no difference: $\epsilon(a) = \epsilon(b) = 0$.

A further condition is that the derivative of the difference function, ϵ also contributes to the function norm used, otherwise non-differentiable functions could also be classified as nearby to a given smooth solution for an orbit stemming from classical physics. We call the nearby function, $f + \epsilon$ as the varied pendant of the function f, frequently also denoted as $f + \delta f$.

This construction leads us to the determination of the variation and the functional derivative of an $I[f]$ functional. The (first) variation of a functional is given as

---

[2] Certainly other equivalent norms can be used, too.

$$\delta I = I[f + \epsilon] - I[f] = \int_a^b dx \, \epsilon(x) \, D(x) + O\left(\|\epsilon\|\right). \tag{1.3}$$

The notation O is the "small ordo", denoting such terms which tend to zero even if divided by their argument. The coefficient function in the integrand, first order in the ϵ difference function, D, is the functional derivative of I. Using the traditional notation

$$D = \frac{\delta I}{\delta f}. \tag{1.4}$$

The operation of the derivative for functionals can be used for searching for extreme values of the functional—in analogy to the world of functions. The condition $\delta I = 0$ is fulfilled only if $D(x) = 0$, except some possible zeroth of the function ϵ. The latter must be restricted to a set of zero measure inside the interval $[a, b]$. In the practice the used test functions, $\epsilon(x)$, are nowhere zero. For an arbitrary ϵ, on the other hand, this means that at the extremum of the functional the function emerging from $D(x) = 0$ is the sought one. $D(x)$ is identically zero for the whole interval in this case, not only in some points x_i.

1.2.3 Higher Functional Derivative

The operation of functional derivative connects a functional with a function, so it leaves the starting set. The repetition of this operation therefore leads to further sets.[3] The interpretation of higher functional derivatives is less straightforward than the derivatives of functions. It is simpler to consider the higher variation. The second variation of a functional is the variation of the first variation:

$$\delta^2 I[f] = \delta I[f] - \delta I[f - \epsilon] = I[f + \epsilon] + I[f - \epsilon] - 2I[f]. \tag{1.5}$$

Such expressions can be viewed as double integrals:

$$\delta^2 I = \int_a^b dx \int_a^b dy \, \epsilon(x) \, M(x, y) \, \epsilon(y) + O\left(\|\epsilon\|^2\right). \tag{1.6}$$

In this expression the two-variable function, $M(x, y)$ is an integration kernel. This, familiar from the theory of Green functions, is not a function but in the general case a *distribution*. Frequently such a kernel solves a differential equation with a general source term, in this way it can be viewed as an inverse of an expression involving differential operators acting on functions. This happens, however, only

---

[3] In field theory the n-point functions are gained from the path integral this way.

if homogeneity is fulfilled, i.e. the kernel is in fact a one-variable function of the difference of its arguments, $M(x, y) = M(x - y)$. Otherwise the kernel can solve an integro-differential equation.

Let a differential equation be given by a polynomial of the operation of derivation against the variable x,

$$P\left(\frac{d}{dx}\right) F(x) = J(x). \tag{1.7}$$

The general solution can be given as

$$F(x) = \int G(x, y) J(y) dy, \tag{1.8}$$

with $G(x, y)$ being the associated Green function. It is a specific case to solve the trivial equation, $F(x) = J(x)$, in this case $P = 1$ identically. The corresponding Green function is not a "function", but a mathematical distribution, the so called *Dirac-delta*. By definition its action under the integral is identical in the sense of

$$F(x) = \int \delta(x - y) J(y) dy = J(x) \tag{1.9}$$

for all possible $J(x)$ source functions. This property helps to obtain the inverse of a Green function. Considering

$$\int K(x, z) G(z, y) dz = \delta(x - y) \tag{1.10}$$

The kernel $K(x, y) = G^{-1}(x, y)$ is to be determined. Based on Eq. (1.8) the following differential equation determines the Green function

$$P\left(\frac{d}{dx}\right) G(x, y) = \delta(x - y). \tag{1.11}$$

In this way it is indeed

$$P\left(\frac{d}{dx}\right) F(x) = \int P\left(\frac{d}{dx}\right) G(x, y) J(y) dy$$

$$= \int \delta(x - y) J(y) dy = J(x). \tag{1.12}$$

A comparison between the Eqs. (1.10) and (1.11) reveals the connection between the polynomial differential operator, $P\left(\frac{d}{dx}\right)$ and the integration kernel $K(x, y)$. Both expressions are the inverse of the Green function.

Seeking for an optimum or extremum raises the question of the stability of the found function or orbit. Similarly to the functional analysis $\delta^2 I > 0$ describes a minimum, and $\delta^2 I < 0$ a maximum. In the case $\delta^2 I = 0$ the nature of the extremum remains indeterminate. The kernels can be analyzed by projecting them into a given basis consisting of functions, $h_i(x)$, for example a Fourier analysis, which reduces the stability investigation to the study of the spectrum of an infinite matrix M_{ij}. In benevolent cases the subspace belonging to zero eigenvalues can be factorized out, and the stability of the extremizing solution can be determined by the rest of the eigenvalues and subdeterminants.

Finally the question naturally arises that if one can calculate with derivatives of functionals against functions, whether the inverse operation, the integration is also possible. Such an operation—and its result—is the functional integral. This does not obtain a "primitive" functional to a given function, assumed to be the result of a functional derivation, but it rather resembles the definite Riemann integral, approximated by ever finer trapezoid contributions: it sums over a functional for all possible function values. In this way such a procedure maps a number to the starting function. Or in more general cases, when containing several functions as arguments, and not integrating over all functions as variables, the result is another functional just having the remaining functions as argument:

$$
Z = \int \mathcal{D}f \; I[f] = \lim_{N \to \infty} \left(\prod_{i=1}^{N} \int df_i \right) I[f(x_i)]. \tag{1.13}
$$

Based on a finite sampling of N possible function values, we consider a set of $f_i = f(x_i)$-s. The $I[f]$ functional is approximated then by an N-variable function, $I[f(x_i)] = I(f_1, f_2, \ldots, f_N)$, and N-fold integrations are calculated. The functional integral is defined as a limit of such calculations for $N \to \infty$.

1.3 A Simple Exercise

Several methods are known for determining extremes of functions, and the strategy is similar in using variational principles. The vanishing functional derivative aka the variation, the complete square form in quadratic cases, investigation of limits of inequalities, or arguing with symmetry, all occur for variational problems, too. In order to demonstrate these basic strategies we solve a very simple problem with several of these methods in the followings. The use of a functional will be the last in this queue (Fig. 1.1).

We seek among right triangles the one with maximal area. Let the length of its span be c, the sides a and b. The area is $t = ab/2$ and the Pythagoras formula, $c^2 = a^2 + b^2$ holds.

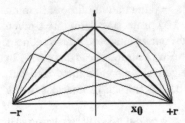 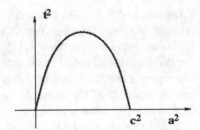

Fig. 1.1 A simple problem: we seek among right triangles with a fixed span those which have a maximal area. Based on the symmetry revealed in the Thales circle (left) this is the case for equal sides ($a = b$), this being the highest of all triangles with the same span. The area squared as a function of the square of one side is shown in the right part of this figure

1.3.1 Completion to a Full Square

The problem is equivalent to obtaining the maximum of $4t^2 = a^2b^2$ with given $a^2 + b^2 = c^2$. The expression $a^2(c^2 - a^2)$ is easily transformed into a full square, $c^2/4 - (c^2/2 - a^2)^2$. This and so the area is then maximal if the sides are equal, $a^2 = c^2/2 = b^2$.

1.3.2 Limit of an Inequality

Twice the area, $2t = ab$ can be interpreted as a geometrical mean, being always less or equal to the arithmetic mean:

$$ab = \sqrt{a^2b^2} \le (a^2 + b^2)/2 = c^2/2.$$

The limit of the inequality above is for equal side squares, $a^2 = b^2$, equivalently at equal sides $a = b$.

1.3.3 Extremum of a Function

Ignoring the geometrical nature of the problem one may try more general methods. Seeking for the maximum of the function $f(a) = 2t = a\sqrt{c^2 - a^2}$, the first derivative vanishes at the maximum:

$$f'(a) = \sqrt{c^2 - a^2} - a^2/\sqrt{c^2 - a^2} = 0.$$

From this condition $a^2 = c^2 - a^2 = b^2$ is given, the known result. The novelty here is that this method is more automatic and works for a number of other problems, too.

1.3.4 Variation of a Functional

One may analyse the above problem using the notion of a functional, too, even if it is not compulsory. The area of a triangle can be calculated as the integral of a function containing two linear pieces. This is a real functional. Denoting the radius of the Thales circle drawn around the triangle by r and putting the coordinate system to its center, the integration runs over the interval $[-r, +r]$, with $r = c/2$ fixed. In this way we are looking for the maximum of the integral

$$I[f] = \int_{-r}^{+r} f(x)dx.$$

By this move we have generalized the problem, we may seek for the maximal area under any curve described by an $f(x)$ function. Even the condition of the fixed span can be changed, e.g. to a given length of the curve. In our case the coordinates of the peak diametrical to the span are x_0 and $f_0 = f(x_0)$. Now the integral $I[f]$ consists of two linear piece contributions and its value is given by

$$I[f] = rf_0 = r\sqrt{r^2 - x_0^2}.$$

This result is an ordinary function of x_0, its maximal value is at $x_0 = 0$ being $I_{max} = r^2$. This also leads to $a = b$.

1.4 A Somewhat More Involved Exercise

In our next problem we seek for that curve of given length which closes a maximal area with the horizontal line. We show, using functional technique, that such a curve must be an arc of a circle.

Let us consider a symmetric arrangement: the curve spans between $x = -a$ and $x = +a$ on the line, and is symmetric to the y axis. Its length then is

$$L = \int_{-a}^{+a} \sqrt{1 + f'(x)^2}\, dx, \tag{1.14}$$

as long as the function $y = f(x)$ describes the curve. Here $f'(x)$ is the derivative of the function f with respect to x. The area to be maximized is the definite integral

$$T = \int\limits_{-a}^{+a} f(x)\, dx. \tag{1.15}$$

Maximizing T with fixed L can be done by using a μ Lagrange multiplier in the functional problem:

$$T - \mu\, L = \text{max.} \tag{1.16}$$

We use the variation of $f(x)$ vanishing in the touch points, $\delta f(x)$:

$$\delta T - \mu\, \delta L = \int\limits_{-a}^{+a} \left(\delta f - \mu\, \frac{f'(x)}{\sqrt{1 + f'(x)^2}}\, \delta f'(x) \right) dx. \tag{1.17}$$

Noting now that $\delta f'(x) = \frac{d}{dx}\delta f(x)$, we perform an integration by parts in the second term. The coefficient of $\delta f(x)$ under the integral is the functional derivative to be set to zero:

$$\frac{\delta}{\delta f}\, (T - \mu\, L) = 1 + \frac{d}{dx}\left(\mu\, \frac{f'(x)}{\sqrt{1 + f'(x)^2}} \right) = 0. \tag{1.18}$$

This is a second order ordinary differential equation with the general solution

$$\frac{f'(x)}{\sqrt{1 + f'(x)^2}} = -\frac{x}{\mu} + C. \tag{1.19}$$

Since due to the symmetry $f'(0) = 0$, the constant $C = 0$ vanishes. Now f' can be expressed as

$$f'(x) = \frac{df}{dx} = \pm\frac{x}{\mu^2 - x^2}. \tag{1.20}$$

Here the relation $\mu^2 \geq a^2$ must hold. Furthermore, since $f(x)$ is rising for negative x and falling for positive x, the negative sign is to be chosen. The solution of the Eq. (1.20) reads as

$$f(x) = f(0) - \int\limits_{0}^{x} \frac{t}{\mu^2 - t^2}\, dt = f(0) + \sqrt{\mu^2 - x^2} - \mu. \tag{1.21}$$

Owing to $f(\pm a) = 0$ one determines $f(0)$ and obtains the final result

$$f(x) = \sqrt{\mu^2 - x^2} - \sqrt{\mu^2 - a^2}. \tag{1.22}$$

This is an equation for an arc of a circle.

Finally the Lagrange multiplier, μ, can be related to the more physical quantities of total length, L, and area T as follows

$$T = \mu^2 \text{ arc sin } \frac{a}{\mu} - a\sqrt{\mu^2 - a^2}, \tag{1.23}$$

and the length,

$$L = 2\mu \text{ arc sin } \frac{a}{\mu}. \tag{1.24}$$

Chapter 2
Mechanics: Geometry of Orbits

It is customary by now to date the birth of modern physics to the dawn of the seventeenth century. Walking in steps of Kepler, Galilei and Newton the *mechanical* worldview develops, and this serves as an idol for other disciplines till date. Electrodynamics at the end of the nineteenth century was interpreted with help of a mechanism transporting fluids and consisting of valves and taps by some contemporary authors, based on the natural existence of a carrying medium of the electric fluid, i.e. the ether. Also the theoretical separating line, quantum mechanics, carries in its name that ideal. Moreover time to time there are public discussions about economical, political and social mechanisms.

In the Anglo-Saxon world this process is used to be regarded as the birth of the "scientific method", which selects out the analysis of causal connections as a noble and effective method in aiming at improvements in our life, unifying the related activity in theory building, discussing an experimental proof in a system of rules of thinking. This basic relation, the causality, is not only a local, step by step principle. Whole causal chains or a contiguous network of those, caused by mutual forces and the changes induced by them, in the mechanics form a complex picture as a whole, and is—although technically involved—always calculable. After clarifying what changes (the state of motion) and how (e.g. following Newton's second law), the noblest challenge in physics what remains is to investigate the nature of forces and the causes behind their appearance.

At the same time it has been gradually revealed that not only the links in the causal chain but also the beginning and the end of the whole chain, initial and boundary conditions play an important role in realizing the basic laws of physics, too. The view based on variational principles comprises these in a single principle, i.e. the local action and the global boundary condition. In this theoretical approach the role of symmetry is ever more emerging; initially just as simplifying ease in understanding the world, but later more and more as an aesthetic category, as a fundamental principle in exploring physical laws in its own right.

© The Author(s), under exclusive license to Springer Nature Switzerland AG 2023   17
T. S. Biró, *Variational Principles in Physics*,
SpringerBriefs in Physics,
https://doi.org/10.1007/978-3-031-27876-1_2

2.1 The Principle of Virtual Work

We start our discussion with a principle determining static mechanical equilibrium positions, namely with the principle of virtual work. Its origin traces back to the 1600 years, its first publication is probably due to Jacob Bernoulli. It states essentially that in an equilibrium situation the forces are so, that for a little imagined movement out of this state, the total work of all the forces is negligible to the same order. In order to formulate this principle precisely one obviously needs calculus with infinitesimal quantities, but it can be used also on the semi-quantitative level intuitively in solving physics problems:

$$\sum_i \mathbf{F}_i \cdot \Delta\mathbf{x}_i = O\left(\|\Delta x\|\right) \tag{2.1}$$

Why does one need more than Newton's third law, assuming the balance between forces? Because the magnitudes of the so called coercive forces are not known, only their direction. A good example is a massive body moving on a slope with a given shape. The direction of the coercive force in a given point is orthogonal to the solid surface, but its magnitude is just as big as ensuring that the movement of the body alongside the slope surface happens. Not a dynamics information, but the geometrical nature of the restriction determines the magnitude of such a force. It does it independently of the material properties of the heavy body and the slope, yet only until a certain limit where either the mass or the slope or both would become too elastic or even fluid. Then dissipation, memory effects and turbulence would influence the motion in a more complicated manner.

The principle of virtual work derives the determination of a static mechanical equilibrium by using the geometry of restrictions by surfaces. Let us review this procedure on the basis of an example of a simple mechanical system.

2.1.1 Halt on the Slope …

Figure 2.1 demonstrates the problem of finding equilibrium on a slope. The left side shows the balance of physical forces, on the right side we have used an inherent symmetry in the situation. It is clear that the chain cannot turn right or left autonomously, without an external action. Furthermore the hanging part obviously balances itself. What remains is the weight proportional to the height of the slope, that is balanced by another weight proportional to the length (span) of the slope. This solution was obtained without using formulas from algebra, purely based on the geometry of arrangement.

Here we shall not discuss methods involving detailed searches of forces, this can be found in several high school textbooks. Instead we turn to the problem solving based on the principle of virtual work, PVW. A consequence of the PVW is that in

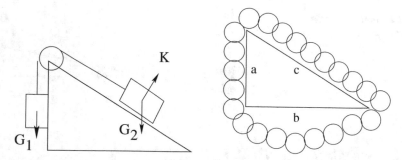

Fig. 2.1 The problem of balance on a slope based on the view of forces (left) and on a symmetric arrangement of weights (right)

gravitational field a small vertical movement of both weights belongs to zero work to the leading order:

$$G_1 \Delta y_1 + G_2 \Delta y_2 = 0. \tag{2.2}$$

Of course, the weight G_2 moves alongside the slope surface, its total $\Delta \ell$ displacement has a vertical component Δy_2. The tightness of the connecting cord and its "inability to be stretched" delivers a pure geometrical condition, not involving the weights:

$$\Delta y_1 + \Delta \ell = 0. \tag{2.3}$$

Finally a rule for similar triangles known already from the classical antique connects the ratio of displacements to sizes of the slope:

$$\frac{\Delta \ell}{\Delta y_2} = \frac{c}{a}, \tag{2.4}$$

with c being the span or length of the slope and a its vertical side, i.e. its height. These three information leads now easily to the "rule of slope", obtained directly from symmetry arguments above:

$$G_1 : G_2 = a : c \tag{2.5}$$

is the condition for equilibrium. In a state near to this equilibrium only a small force suffices to hive heavy weights, especially if the inclination of the slope and with that the ratio $a : c$ is small enough. Using this knowledge gigantic pyramids had been erected.

Fig. 2.2 Equilibrium for
two arm bascule

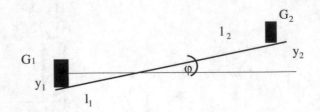

2.1.2 Bascule in Equilibrium

In a similar way one can use the principle of virtual work (PVW) for describing the equilibrium of a bascule. According to the PVW, when the weight G_1 is lifted by Δy_1 the counterweight G_2 sinks by Δy_2, the balance is given by

$$G_1 \Delta y_1 - G_2 \Delta y_2 = 0. \tag{2.6}$$

The geometry of the constraint is based on the rigidity of the arms: the angles at an imagined small rotation around an axis spearing through the point of support are equal:

$$\tan \varphi = \frac{\Delta y_1}{\ell_1} = \frac{\Delta y_2}{\ell_2}. \tag{2.7}$$

From these two equations the "bascule rule" emerges (Fig. 2.2)

$$G_1 \ell_1 = G_2 \ell_2. \tag{2.8}$$

2.2 D'Alambert Principle

The D'Alambert principle in modern view is nothing else than the principle of virtual work if inertial forces due to acceleration are taken into account. These inertial forces appear in accelerating coordinate systems, common in all of them, that their magnitude is proportional to the inertial mass. Such are the centrifugal force felt in a vehicle taking a sudden turn, or the Coriolis force acting on moving masses in a rotating system, also responsible for huge cyclones in Earth's atmosphere. In some classical western movies a glass of whiskey is shown as it slides on the table: since the surface of the liquid is orthogonal to the sum of forces, this direction is no more horizontal, unlike for a standing glass with whiskey. (See Fig. 2.3).

The principle of general relativity states that the laws of physics are independent from the accelerating motion of observers, those can even be automatic devices, detector systems without a human. Therefore all accelerating motion can be described as a static equilibrium problem inside a comover coordinate system. General relativity

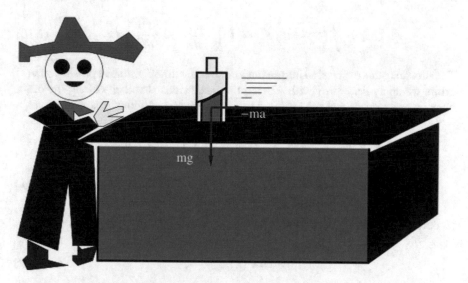

Fig. 2.3 Illustration to the interpretation of the D'Alambert principle as a PVW: the inertial force $-ma$ has to be taken into account when explaining the surface of liquid in an accelerating glass

tells us something about gravitation, because the gravitating mass, occurring in the Newtonian law of gravity, and the inertial mass, occurring in Newton's second law, are equal, as it was justified first by experiments run by the Hungarian Baron Roland Eötvös.

However, for the contemporaries the D'Alambert principle appeared as an autonomous principle on its own. It served to describe the motion of mass points or center of mass points in generally moving systems. If all forces—including the inertial "pseudo" forces—are taken into account in a calculation of the work for a virtual displacement, the vanishing of this total work to leading order constitutes D'Alambert's principle:

$$\sum_i (F_i - m\ddot{x}_i)\, \Delta x_i = 0. \tag{2.9}$$

Some regard this formula as a consequence of Newton's second law, but in that derivation not only the sum of all terms vanishes, but also each term separately. This situation is symmetric: for taking the D'Alambert principle (DAP) for arbitrary virtual displacements, Δx_i, is already equivalent with Newton's law. In this sense DAP is a (discrete) variational principle.

For conservative force fields the force is the negative gradient of a potential energy, for one dimensional motion $F = -dV/dx$. As an illustration we show here that in this case the DAP expresses the constancy of the sum of kinetic and potential energy. For this purpose we consider infinitesimal virtual displacements and turn the sum into an integral

$$D = \int (F - m\ddot{x})\, dx = \int \left(-\frac{dV}{dx} - m\frac{d^2x}{dt^2} \right) dx. \qquad (2.10)$$

Transforming this expression into an integral over the time coordinate passing on $x(t)$ orbits we apply $dx = v\, dt$ with v being the instanteneous velocity along the orbit. Now one recognizes that the DAP led to a time integral of a total time derivative:

$$D = \int \left(-\frac{dV}{dx} - m\frac{d^2x}{dt^2} \right) \frac{dx}{dt}\, dt = \int \left(-\frac{dV}{dt} - mv\frac{dv}{dt} \right) dt$$

$$= -\int \frac{d}{dt} \left(V + \frac{mv^2}{2} \right) dt. \qquad (2.11)$$

According to DAP we have $D = 0$, and it follows that the quantity,

$$E = V(x) + \frac{mv^2}{2}, \qquad (2.12)$$

the sum of kinetic and potential energy is constant in time.

A particular case is the free motion of a mass point when the potential energy vanishes, $V = 0$. The kinetic energy of an arbitrary system of freely moving masses m_i with velocities v_i becomes

$$T = \frac{1}{2} \sum_{i=1}^{N} m_i \mathbf{v}_i^2, \qquad (2.13)$$

constant in time, therefore one may describe that motion in the $3N$-dimensional configurational space by a differential arc length square, weighted by masses, as follows:

$$ds^2 = 2T\, dt^2 = \sum_{i=1}^{N} m_i(dx_i^2 + dy_i^2 + dz_i^2). \qquad (2.14)$$

This is the differential geometric description of the free motion. The use of generalized coordinates (like angles and radii) leads to a more powerful formula:

$$ds^2 = \sum_{i,k} a_{ik}(q)dq^i dq^k. \qquad (2.15)$$

This form will be the foundation for the Mapertuis approach, seeking for the shortest path.

In the followings we use two examples for applying the DAP: free motion on a circle and the swings of a pendulum.

2.2.1 Free Circular Motion

A potential-free motion on a circle can be treated in two basic ways. According to the direct method one selects that generalized coordinate in terms of which the real motion is the simplest. For this we have to know a lot about the solution orbit in advance. We also may use spatial Cartesian coordinates and then relate them due to the form of the orbit.

We choose the Cartesian coordinates x and y in the plane of the circle. Then the square of the arc length differential for free motion is given as

$$ds^2 = 2T dt^2 = m(dx^2 + dy^2). \tag{2.16}$$

On the other hand it is immediately obvious that in polar coordinates the radius of a circle is constant, so the arc length square has a more concise expression in terms of an angle differential:

$$ds^2 = mR^2 d\varphi^2. \tag{2.17}$$

The free motion on a circle is one-dimensional, although not straight, but still smooth: $\varphi(t) = \varphi(0) + \omega t$. By a fixed kinetic energy, T, the angular frequency of the circular motion is $\omega = \sqrt{2T/mR^2}$, and the period is $2\pi/\omega$. The velocity on the orbit has a constant magnitude, $v = R\omega$.

Alternatively, the DAP can be applied blindly not considering any speciality of the circular geometry in the beginning. More precisely the latter will be taken into account in the language of naive, Cartesian coordinates. The DAP confirms

$$m(\ddot{x}dx + \ddot{y}dy) = 0, \tag{2.18}$$

while the equation of a circle in Descartes coordinates is given by $x^2 + y^2 = R^2$, hence

$$xdx + ydy = 0 \tag{2.19}$$

follows. From this $dy = -xdx/y$ is expressed and the arc length square at constant kinetic energy becomes

$$ds^2 = 2T dt^2 = m(1 + x^2/y^2)dx^2 = \frac{mR^2}{R^2 - x^2}dx^2. \tag{2.20}$$

This leads to a first order differential equation

$$dx = v_0\sqrt{1 - \frac{x^2}{R^2}}dt \tag{2.21}$$

featuring the equation of free circular motion in terms of the coordinate x. This differential equation (2.21) is separable and can be solved by a single integration:

$$\int \frac{R\,dx}{\sqrt{R^2 - x^2}} = \int v_0 \, dt. \tag{2.22}$$

From the result of integration one may express x as a function of time,

$$x(t) = R \cos\left(\varphi_0 + \frac{v_0}{R} t\right). \tag{2.23}$$

This result is equivalent with the previous one with the replacement $\omega = v_0/R$.

In the present case the constraint was an algebraic function of the coordinates only; such restrictions in the form $f(q_1, \ldots, q_N) = 0$ are holonomic. For the DAP, however, only the differential form of the constraint, $\sum_i A^i(q) dq_i = 0$ counts. Whenever this form is not integrable, the constraint is not holonomic. A further classification distinguishes time dependent, rheonomic, constraints and time independent, scleronomic, constraint. But this is already like botanics.

2.2.2 Pendulum

The motion of a pendulum in our age appears as a training exercise in high schools. However, in the 16th century it was a very important problem for the humanity. An ancestor of the pendulum clock, Galilei's pendulum, was good for measuring time, better than counting men's own pulse or using a sand glass. Its only disadvantage is that its motion is not uniform over time, since the weight moves in a gravitational field, changing its potential energy, too. A reliable measurement of time became also important for navigation on the sea, because the longitude of a ship's position without knowing the precise time, only based on the instantaneous picture of the night sky could not be determined. No wonder that the British Admiralty announced a special prize for constructing a mechanical clockwork which is beating uniformly in time. A variation of the pendulum which ticks uniformly in time, the solution of the so called isochron problem, is due to Huygens. The practical solution some time later became the mechanical clock with springs and gears.

The swings of the simple physical pendulum can be described in a coordinate system using the suspension point as origin. The y axis is vertical, along that the potential energy changes, whose derivative coincides with $-mg$ force. The DAP is of the following form

$$-mg\,dy - m\ddot{y}\,dy - m\ddot{x}\,dx = 0. \tag{2.24}$$

The geometry of the pendulum, with a tensed cord with constant length, provides the following relation between the coordinates:

$$
\begin{aligned}
y &= L\left(1 - \cos\varphi\right), \\
x &= L \sin\varphi.
\end{aligned} \tag{2.25}
$$

Since only the actual angle of the cord changes during the motion of a pendulum, the coordinate differentials are related to the angle differential as

$$dy = L \sin \varphi \, d\varphi,$$
$$dx = L \cos \varphi \, d\varphi. \tag{2.26}$$

From this the first and second time derivatives emerge,

$$\dot{y} = L \sin \varphi \, \dot{\varphi}, \ddot{y} = L \cos \varphi \, \dot{\varphi}^2 + L \ddot{\varphi} \sin \varphi,$$
$$\dot{x} = L \cos \varphi \, \dot{\varphi}, \ddot{y} = -L \sin \varphi \, \dot{\varphi}^2 + L \ddot{\varphi} \cos \varphi,$$

$$\tag{2.27}$$

concluding in the DAP part describing the inertial forces:

$$\ddot{x} dx + \ddot{y} dy = L^2 \ddot{\varphi} \, d\varphi. \tag{2.28}$$

In the complete DAP also the work of gravity is taken into account, so we arrive at

$$-mL^2 d\varphi \left(\frac{g}{L} \sin \varphi + \ddot{\varphi} \right) = 0. \tag{2.29}$$

The vanishing of the quantity in the bracket delivers the equation of motion for the pendulum. It is noteworthy that we have obtained this result based only on the inertial forces and the principle of virtual work; no word about the angular momentum has been used.

2.3 The Action Principle

The action principle is the successor of the DAP on the throne of variational principles in mechanics. In fact it can be derived from it. However the point of view is different: so far we have analyzed a sum of little work terms connected to virtual displacements, declared to vanish in the leading order. From now on we shall be looking for an extremum (maximum or minimum) of something.

As the PVW (principle of virtual work) due to Bernoulli regards a sum of imagined works, the $dW = \sum_i F_i dx^i$ one-form functional, we construct another one, the *action functional*, connected to D'Alambert's principle:

$$\sum_i (F_i - m\ddot{x}_i) dx^i = d\dot{S}[x_i]. \tag{2.30}$$

Certainly, since the work done by inertial forces contains the accelerations, such an action must depend on the time derivatives of orbital curve $x_i(t)$, too. Finally

$$S = S\left[x_i(t), \dot{x}_i(t), \ldots, t\right] \tag{2.31}$$

is the general form of the action. Higher than first derivatives of the sought solution curve may appear in problems regarding the shape of elastic rods under weight, but in the mechanics of mass points these will not occur. The explicit time interval dependence is also absent for conservative (non dissipative) systems.

In the presence of conservative force fields, $F_i = -\partial V/\partial x^i$, so the complete action can be written as a time integral of a *Lagrange function*. We write the sum of work and "inertial work" as an infinitesimal change in the action:

$$dS = \int_{t_1}^{t_2} \sum_i \left(F_i - \frac{d}{dt}(mv_i)\right) dx^i dt. \tag{2.32}$$

This expression, after substituting the components of conservative forces by corresponding gradient terms of a potential energy, transmutes into

$$dS = -\int_{t_1}^{t_2} \sum_i \left(\frac{\partial V}{\partial x^i} + m\dot{v}_i\right) dx^i dt. \tag{2.33}$$

Integrating the second term by parts we obtain

$$dS = -\left. mv_i dx^i \right|_{t_1}^{t_2} + \int_{t_1}^{t_2} \sum_i \left(mv_i(\dot{dx}^i) - \frac{\partial V}{\partial x^i} dx^i\right) dt. \tag{2.34}$$

Since on the limits of integration the variations vanish, $dx^i(t_1) = dx^i(t_2) = 0$, exchanging the order of time derivative and orbit variation we recognize the variation of velocity components as $(\dot{dx}^i) = dv^i$. Moreover the sum contains a total time change of the conservative potential energy, $V(x)$. Finally the change of the action functional is given by

$$dS = d\int_{t_1}^{t_2} \left(\sum_i \frac{m}{2}v_i^2 - V(x)\right) dt. \tag{2.35}$$

Formulating in words, the *difference* between kinetic and potential energy is the Lagrange function, whose integral in time is the action functional,

$$S = \int_{t_1}^{t_2} (T - V)\, dt = \int_{t_1}^{t_2} L\, dt. \tag{2.36}$$

The first variation of this action functional set to zero determines the classical equations of motion.

For a general conservative Lagrange function, $L = L(x, \dot{x})$, the variational action principle leads to a general form of equations of motion, to the *Euler-Lagrange equations*. This derivation is widespread, it can be found in all textbooks on theoretical mechanics. We present it here for the sake of completeness.

The variation of the action,

$$\delta S = \delta \int L \, dt = \int \left(\frac{\partial L}{\partial x_i} \delta x_i + \frac{\partial L}{\partial \dot{x}_i} \delta \dot{x}_i \right) dt \tag{2.37}$$

after integration by parts becomes

$$\delta S = \int \left(\frac{\partial L}{\partial x_i} - \frac{d}{dt} \frac{\partial L}{\partial \dot{x}_i} \right) \delta x_i \, dt = 0. \tag{2.38}$$

Since this action-variation must vanish for arbitrary orbit-variations, $\delta x_i(t)$, the expression inside the brackets—being at the same time the first functional derivative of the action functional—is identically zero:

$$\frac{\delta S}{\delta x_i} = \frac{\partial L}{\partial x_i} - \frac{d}{dt} \frac{\partial L}{\partial \dot{x}_i} = 0. \tag{2.39}$$

The orbital velocities, $v_i(t) = \dot{x}_i(t)$ and positions, $x_i(t)$ satisfy the equation of classical motion above. Here the $x_i(t)$ are not necessarily Cartesian coordinates, they are in general real functions of time describing the motion in general terms.

Another view of the variation considers the case when the variation does not vanish at the integration limits (this is sometimes called a "second type" variation), but the Euler–Lagrange equation (2.39) is fulfilled. Then the remaining variation of the action is from the integrated product term by the integration by parts:

$$\delta_{(2)} S = \left. \frac{\partial L}{\partial \dot{x}_i} \delta x_i \right|_{t_1}^{t_2} = 0. \tag{2.40}$$

In case the coordinate variations at the endpoints are nonzero and arbitrary, it can be viewed as a consequence of an infinitesimal coordinate transformation, $x'_i = x_i + \delta x_i(\epsilon^a)$, parameterized by the ϵ^a quantities, during which the action functional is invariant, then the following expression is constant in time:

$$Q^a = \sum_i \frac{\partial L}{\partial \dot{x}_i} \cdot \frac{\partial}{\partial \epsilon^a} \delta x_i. \tag{2.41}$$

Such Q^a conserved quantities are called *Noether charge*. Most known are those belonging to a coordinate shift, $\delta x_i = \epsilon_i$, being the momenta, $P_i = \frac{\partial L}{\partial \dot{x}^i}$ and to a time shift $t' = t + \epsilon$, with the corresponding energy. In the latter case the change in the action by such a transformation can be calculated in two ways leading to the same result: first by adding ϵ-times the total time derivative of the action functional to itself, second by pushing the orbital points towards those belonging to a later time. In this way

$$S' = S + \epsilon L = \int L(x + \epsilon \dot{x}, \dot{x} + \epsilon \ddot{x}) \, dt. \tag{2.42}$$

Collecting terms of order ϵ from this equation we obtain

$$\int \left(\dot{x} \frac{\partial L}{\partial x} + \ddot{x} \frac{\partial L}{\partial \dot{x}} - \frac{dL}{dt} \right) dt = 0. \tag{2.43}$$

Using now the Euler-Lagrange equation (2.39), the expression under the integral becomes a total time derivative, whose integral between the time instants t_1 and t_2 gives the difference

$$H(t_2) - H(t_1) = \int \frac{d}{dt} \left(\dot{x} \frac{\partial L}{\partial \dot{x}} - L \right) = 0. \tag{2.44}$$

The difference is zero, therefore the conserved energy can be expressed from the Lagrange functions. The result is the Hamilton function:

$$H = \sum_i \dot{x}_i \, P^i - L. \tag{2.45}$$

This formula states also that the Hamiltonian is a Legendre transform to the Lagrange function. Here the independent variables are changed from the time derivatives of generalized coordinates, i.e. generalized velocities $v_i = \dot{x}_i$, to the corresponding momentum components, P_i, defined by the symmetry against spatial shifts.

For non conservative dynamical systems, the action explicitly depends on time. In this more general case the partial derivative of the action contributes to the total change, too. The Hamiltonian is not constant in time in this case, but it exactly compensates the change due to the partial time derivative. This statement is comprised in the Hamilton–Jacobi equation:

$$H + \frac{\partial S}{\partial t} = 0. \tag{2.46}$$

Here in the arguments of the Hamiltonian the momenta are replaced by the corresponding gradient components of the action, $P_i = \frac{\partial S}{\partial x^i}$.

2.4 Second Variation of the Free Motion

The annihilation of the first variation of the action delivers those differential equations which describe the classical motion. The second variation hints at the stability of those orbits in the framework of the variational principle. Here we do not deal with the general case, instead for the free motion of a mass point with mass m will be the second variation of the action calculated.

The motion is defined by the following Lagrange function:

$$L = \frac{1}{2} \sum_i m_i \dot{x}_i^2. \tag{2.47}$$

The corresponding Euler–Lagrange equation (2.39),

$$\frac{\partial L}{\partial x_i} - \frac{d}{dt} \frac{\partial L}{\partial \dot{x}_i} = -m_i \ddot{x}_i = 0, \tag{2.48}$$

describes a motion without acceleration. The differential operator associated to the integration kernel in the expression of the second variation of the action, after integration by parts, delivers the coefficient matrix in the quadratic term containing the velocities, $\dot{x}_i$:

$$M_{ij} = -\delta_{ij} \frac{d}{dt} m_i \frac{d}{dt}. \tag{2.49}$$

For time independent masses this is a constant matrix times the operation of the second derivative with respect to the time variable. The eigenvalue spectrum of this matrix presents the stability analysis of the classical solution. The eigenvalues are determined from the vanishing of the determinant

$$\det\left(M_{ij} - \omega^2 \delta_{ij}\right) = 0, \tag{2.50}$$

one has to solve this "secular" equation. In order to obtain this one uses the eigenfunctions of the matrix M_{ij}, i.e.

$$M_{ij} f_j^{(n)}(t) = \omega_n^2 f_i^{(n)}(t). \tag{2.51}$$

The eigenfunctions for the free motion without acceleration are solutions of the following differential equations:

$$-m \frac{d^2}{dt^2} f_i^{(n)}(t) = \omega_n^2 f_i^{(n)}(t). \tag{2.52}$$

The boundary conditions fix the variations at the beginning and at the end of the time-integration, therefore for all eigenfunctions $f_i^{(n)}(t_1) = 0$ and $f_i^{(n)}(t_2) = 0$. Such solutions to the Eq. (2.52) are sine functions:

$$f_i^{(n)}(t) = A_i^{(n)} \sin\left(\frac{\omega_n}{\sqrt{m}} (t - t_1)\right) \qquad (2.53)$$

Satisfying now also the vanishing first variation of the action, the eigenvalues must obey

$$\frac{\omega_n}{\sqrt{m}} (t_2 - t_1) = n\,\pi. \qquad (2.54)$$

The variation of orbits, like a string fixed at its endpoints, produces oscillations. Its frequencies are functions of the length of the time-interval:

$$\omega_n = \frac{\pi\sqrt{m}}{t_2 - t_1} n. \qquad (2.55)$$

The orbit solution found by variation of the action, since all ω_n^2 values are positive, and therefore the frequencies are real, is stable. Formally there is also a mode with zero frequency, but this would describe a variation identically zero in the full time interval. Such a null-variation is usually excluded when solving the variational principle (Fig. 2.4).

An arbitrary variation can be expanded in terms of the above eigenfunctions,

$$\delta x_i(t) = \sum_n c_n f_i^{(n)}(t) \qquad (2.56)$$

Fig. 2.4 The $x(t)$ orbit of uniform inertial motion will be varied by arbitrary $\delta x(t)$, fixed in the strating endpoint. Then the expansion in terms of sine functions is natural

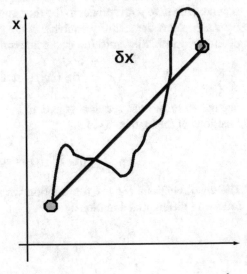

is a periodic function with the period $t_2 - t_1$. Its function norm can be expressed with the help of Fourier coefficients, where a norm of one can be chosen for each mode rescaling the coefficients, c_n, only. Finally

$$\|\delta x_i(t)\|^2 = \int\limits_{t_1}^{t_2} \sum_i \delta x_i^2(t) dt = \frac{1}{2} \sum_{n,i} |c_n|^2 |A_i^{(n)}|^2, \tag{2.57}$$

is the norm of the orbit variation, based on the energy theorem of Fourier.

2.5 Gauss Principle

Carl Friedrich Gauss has enriched the science in mathematics and physics with several novelties, we use even to date. Besides the well-known Gauss–distribution (the bell shape curve) and the—among physicists and mathematicians popular Gauss–Ostrogodskii theorem, such an everyday tool, like the fit of a most probable line onto a number of data points minimizing the sum of the squares of distances from this line, all are related to his name. The Gauss principle in the mechanics is in fact the minimization of deviances from Newton's second law by the Gauss method.

This result can be motivated starting with D'Alambert's principle (DAP). Imagine that during the motion the coordinate functions change a little due to a change in the time:

$$x_i(t + \tau) = x_i(t) + v_i(t)\,\tau + \frac{1}{2}a_i(t)\,\tau^2 + \cdots \tag{2.58}$$

Since in the mechanics the initial position and velocity can be fixed, the leading order in the variation during this short τ time can be traced back to the variation of acceleration:

$$\delta x_i(t + \tau) = \frac{1}{2}\delta a_i(t)\,\tau^2 \tag{2.59}$$

being the difference between the original and the varied motion. Writing the DAP both at t and at $t + \tau$, their difference will contain the variation of acceleration only:

$$\frac{1}{2}\tau^2 \sum_i (F_i - m_i a_i)\,\delta a_i = 0. \tag{2.60}$$

Since the forces F_i are given by the natural laws and by the specialities of the arrangement under scrutiny, these cannot be varied: $\delta F_i = 0$. Therefore the variation of the acceleration is essentially the variation of the whole expression in the bracket, $\delta a_i = -\delta(F_i - m_i a_i)/m_i$, so Eq. (2.60) cab be written as a total variation being zero:

$$\frac{1}{2}\tau^2\delta\left(\sum_i \frac{1}{2m_i}(F_i - m_i a_i)^2\right) = 0. \tag{2.61}$$

This is equivalent with the condition for an extremum of the varied sum—logically its minimum:

$$G = \sum_i \frac{1}{2m_i}(F_i - m_i a_i)^2 = \min. \tag{2.62}$$

The above is the Gauss–principle, it minimizes the squares of deviances from Newton's second law weighted by inverses of the masses. Although in the absolute minimum of this expression Newton's law is fulfilled for all particles with mass m_i separately, for describing the motion less is enough. The Gauss–principle is best used if certain forces are unknown, and rather geometric constraints cause them (Fig. 2.5).

As an example we discuss the motion on a flat slope. The equation determining the general constraint surface is of the form $z = f(x, y)$, for a flat slope this function is multilinear, $z = ax + by + c$. Gravity is a known force, acting in the z direction, $F_z = -mg$. Due to the constraint the z-component of the velocity and acceleration are not independent from the other two components: derivatives of the constraint give

$$\dot{z} = f_x\dot{x} + f_y\dot{y},$$
$$\ddot{z} = f_x\ddot{x} + f_y\ddot{y} + f_{xx}\dot{x}^2 + (f_{xy} + f_{yx})\dot{x}\dot{y} + f_{yy}\dot{y}^2. \tag{2.63}$$

Here the lower indices to the function f denote partial derivatives with respect to the noted variable. The Gauss–principle for a single mass is reduced to the problem of minimization of the expression below:

$$2mG = (F_x - m\ddot{x})^2 + (F_y - m\ddot{y})^2 + (F_z - m\ddot{z})^2 = \min. \tag{2.64}$$

Fig. 2.5 A motion in 3D restricted by a 2D surface under the influence of vertical gravity force is a typical application area for the Gauss principle

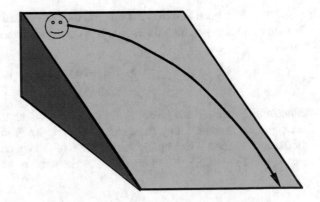

This quantity is to be varied against $\ddot{x}$ and $\ddot{y}$, a possible variation against $\ddot{z}$ would not contain additional information:

$$\frac{\delta}{\delta \ddot{x}}(F_z - m\ddot{z}) = -m\frac{\delta \ddot{z}}{\delta \ddot{x}} = -mf_x,$$

$$\frac{\delta}{\delta \ddot{y}}(F_z - m\ddot{z}) = -m\frac{\delta \ddot{z}}{\delta \ddot{y}} = -mf_y. \tag{2.65}$$

These two independent variations of the Gauss–principle result in

$$2m\frac{\delta G}{\delta \ddot{x}} = 2(F_x - m\ddot{x})(-m) + 2(F_z - m\ddot{z})(-mf_x) = 0,$$

$$2m\frac{\delta G}{\delta \ddot{y}} = 2(F_y - m\ddot{y})(-m) + 2(F_z - m\ddot{z})(-mf_y) = 0. \tag{2.66}$$

The equations of motion below follow from it,

$$m(\ddot{x} + f_x\ddot{z}) = (F_x + f_x F_z),$$
$$m(\ddot{y} + f_y\ddot{z}) = (F_y + f_y F_z). \tag{2.67}$$

What is still needed is a substitution of $\ddot{z}$ from the Eq. (2.63), in the general case leading to rather involved formulas. However, there exists a combination of the above two lines which is independent from $\ddot{z}$:

$$\dot{L}_z = m(f_y\ddot{x} - f_x\ddot{y}) = f_y F_x - f_x F_y. \tag{2.68}$$

Finally using the formula for a flat slope, $z = ax + by + c$, and the known components of gravitational force, $F_x = 0$, $F_y = 0$, $F_z = -mg$, the equations of motion in (2.67) resemble a much simpler form:

$$\frac{1}{a}\ddot{x} + a\ddot{x} + b\ddot{y} = -g,$$

$$\frac{1}{b}\ddot{y} + a\ddot{x} + b\ddot{y} = -g. \tag{2.69}$$

This in turn delivers the following result for the acceleration components:

$$\ddot{x} = -g\frac{u}{1 + a^2 + b^2},$$

$$\ddot{y} = -g\frac{b}{1 + a^2 + b^2},$$

$$\ddot{z} = -g\frac{a^2 + b^2}{1 + a^2 + b^2}. \tag{2.70}$$

The absolute value of the acceleration vector can be expressed from the above formula for its components as being

$$|a| = \sqrt{\ddot{x}^2 + \ddot{y}^2 + \ddot{z}^2} = g\sqrt{\frac{a^2 + b^2}{1 + a^2 + b^2}} = g \sin \alpha, \qquad (2.71)$$

with α being the angle of the slope to the horizontal plane. Exactly this reduction of the gravitational acceleration, $g \sin \alpha$ led Galileo Galilei to make experiments on slopes, because in this way the free fall can be examined in a time zoom.

2.6 The Method of Lagrange Multipliers

It is advantageous to follow variational principles instead of directly applying Newton's equation whenever we consider geometric constraints on the orbits of motion. But most problems are such! The Lagrange method aims to follow all constraints systematically. This motivation moved Lagrange to suggest the use of generalized coordinates; those are not only naturally fit to some symmetries inherent in the problem, but also follow the independent degrees of freedom for possible motions. Unfortunately it is in the least cases obvious that which generalized coordinates belong to the "pure" degrees of freedom. Those rare exemptions are treated habitually in textbooks.

In most cases it is comfortable to formulate both the variational principle and the constraints in terms of a naive, e.g. Cartesian coordinate system. Then the mathematical task is to vary the relation $F(x_1, x_2, \ldots, x_n) = $ extremum constrained by some $f(x_1, x_2, \ldots, x_n) = 0$ condition. The naive method expresses one of the coordinates, say x_n, from the constraint and considers the quantity to be varied after substituting this result:

$$\tilde{F}(x_1, x_2, \ldots, x_{n-1}) = F(x_1, x_2, \ldots, x_n(x_1, x_2, \ldots, x_{n-1})) = \text{extremum.} \quad (2.72)$$

Lagrange suggested instead to think differently: the constraints are valid both for the original and for the varied orbits, therefore both F's and f's first variation vanishes at the solution:

$$\delta F = \sum_i \frac{\partial F}{\partial x_i} \delta x_i = 0,$$

$$\delta f = \sum_i \frac{\partial f}{\partial x_i} \delta x_i = 0. \qquad (2.73)$$

Furthermore an arbitrary linear combination of the above two equations is also zero; the admixture coefficient is the Lagrange multiplier. λ:

$$\delta \left(F + \lambda f \right) = 0. \tag{2.74}$$

The strategy with Lagrange multipliers is as follows: (i) from one of the variations above, say the n-th, we obtain λ, and then ii) the rest $n - 1$ equations are treated as equations of motion for the coordinates $(x_1, x_2, \ldots, x_{n-1})$. Certainly beyond that the constraint $f = 0$ is also to be used at the end, but this type of equations are often simpler than those containing the derivatives of the original functions, like when using the Gauss-principle in the previous section.

In case of several constraints the Lagrange method leads to the modified variational principle

$$\bar{F} = F + \sum_{i=1}^{m} \lambda_i \, f_i = \text{extremum}, \tag{2.75}$$

treated as a function of $n + m$ generalized coordinates, $(\lambda_1, \lambda_2, \ldots, \lambda_m, x_1, x_2, \ldots, x_n)$. Variations against the original x_i coordinates delivers n equations containing the Lagrange multipliers, while the variation against the extra coordinates λ_i gives back the m constraints:

$$\frac{\delta \bar{F}}{\delta x_k} = \frac{\delta F}{\delta x_k} + \sum_{i=1}^{m} \lambda_i \frac{\delta f_i}{\delta x_k} = 0,$$

$$\frac{\delta \bar{F}}{\delta \lambda_i} = f_i = 0. \tag{2.76}$$

As an example we determine the equilibrium shape of a chain with homogeneous mass distribution hung at its two ends, the chain-curve (see Fig. 2.6). We divide the continuous chain to small pieces. The center of mass coordinates of a given element be x_k and y_k. The lengths of the chain elements are constant, this gives N constraints among $2N$ coordinates:

Fig. 2.6 Horizontally fixed chain under gravity forms in equilibrium the special chain-curve. Figure credit: ©Kamel15/Wikipedia. Reproduced under CC BY-SA 3.0 license

$$\ell_k^2 = (x_{k+1} - x_k)^2 + (y_{k+1} - y_k)^2. \tag{2.77}$$

The potential energy of the arrangement is to be minimized, that depends on the vertical positions of the chain elements:

$$V = g \sum_{k=0}^{N-1} m_k \bar{y}_k = \frac{g\sigma}{2} \sum_{k=0}^{N-1} (y_k + y_{k+1}) \ell_k, \tag{2.78}$$

with g being the gravitational acceleration and σ the constant mass density along the chain. It is purposeful to choose the Lagrange multipliers $\lambda_k/2$ and to minimize

$$\bar{V} = V + \sum_{k=0}^{N-1} \frac{\lambda_k}{2} \ell_k^2 \tag{2.79}$$

with above constraints. Substituting the expression to be varied becomes

$$\bar{V} = \frac{1}{2} \sum_{k=0}^{N-1} \left((y_k + y_{k+1}) g\sigma \ell_k + \lambda_k (x_{k+1} - x_k)^2 + \lambda_k (y_{k+1} - y_k)^2 - \lambda_k \ell_k^2 \right). \tag{2.80}$$

Variation with respect to the x_k coordinates gives

$$\frac{\delta \bar{V}}{\delta x_k} = \lambda_k (x_{k+1} - x_k) - \lambda_{k-1} (x_k - x_{k-1}) = 0. \tag{2.81}$$

This can be written as a recursive formula by introducing the differences in the x_k series, $\Delta x_k = x_{k+1} - x_k$, as

$$\lambda_k \, \Delta x_k = \lambda_{k-1} \, \Delta x_{k-1}. \tag{2.82}$$

Its solution is $\lambda_k = c/\Delta x_k$, with an arbitrary constant, c.

The result of the variation of the x_k coordinates deliver

$$\frac{\delta \bar{V}}{\delta y_k} = \frac{g\sigma}{2}(\ell_k + \ell_{k-1}) - \lambda_k \, \Delta y_k + \lambda_{k-1} \, \Delta y_{k-1} = 0. \tag{2.83}$$

Substituting the form for λ_k obtained earlier we get

$$c \left(\frac{\Delta y_k}{\Delta x_k} - \frac{\Delta y_{k-1}}{\Delta x_{k-1}} \right) = \frac{g\sigma}{2} (\ell_k + \ell_{k-1}). \tag{2.84}$$

It is clear that a useful choice for the constant c is $c = g\sigma(\ell_k + \ell_{k-1})/2$, meaning a uniform length division of the chain to elements. In the limit of continuity the

Fig. 2.7 Taking into account restrictive conditions with a quadratic potential. The smaller the parameter ϵ the sharper the constraint

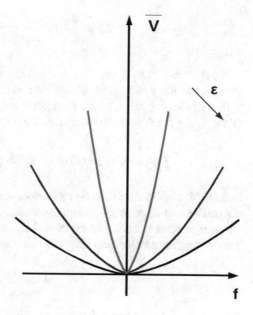

coordinate differences mute into differentials along continuous curves, so finally we arrive at the following *chain equation*:

$$\frac{d}{ds}\left(\frac{dy}{dx}\right) = 1. \tag{2.85}$$

Finally one may be interested in the shape of the curve in a non-parametric description, as an $y(x)$ function. Then the above equation appears as a second order differential equation,

$$\frac{1}{\sqrt{1 + y'(x)^2}} \, y'' = 1. \tag{2.86}$$

After a little re-arrangement, $(y'')^2 - (y')^2 = 1$, and it is straightforward to see that the solution chain-curve is cosine hyperbolic, $y = \cosh(x)$ if the x coordinate origin is at the deepest point of the curve (Fig. 2.7).

To the Lagrange method one may attach a physical picture, too. The terms added in the potential energy appear in the Lagrange function as subtraction,

$$\bar{L} = L - \sum_{i_1}^{m} \lambda_i f_i. \tag{2.87}$$

The negative gradient of this completed potential gives exactly the coercive force components:

$$K_i = \frac{\partial \bar{L}}{\partial x_i} - \frac{\partial L}{\partial x_i} = -\sum_{j_1}^{m} \lambda_j \frac{\partial f_j}{\partial x_i}, \tag{2.88}$$

while the constraints $f_j(x_1, \ldots, x_n) = 0$ are satisfied. Imagine now a cost factor in the space of the f_j components, acting analogously to the potential energy. Starting with a single constraint, which does not change the essence of the logic, we expand this quantity in a Taylor series around the point $f = 0$:

$$V_1 = \Phi(f) = \Phi(0) + \Phi'(0) f + \frac{1}{2}\Phi''(0) f^2 + \cdots \tag{2.89}$$

Here $\Phi(0)$ can be chosen to be zero, since a constant in the energy landscape does not matter, and we omit the linear term assuming $\Phi'(0) = 0$. That can be achieved by choosing a corresponding linear combination of constraint as the "real" constraint. The leading term in the cost factor in this way is the quadratic one:

$$V_1 = \frac{1}{2\epsilon} f^2. \tag{2.90}$$

One may consider this form also as additive while applying the Lagrange method; the condition $f^2/2 = 0$ is equivalent to $f = 0$. The coercive forces, determined from these two—quadratic and linear—approaches, must be equal,

$$K_i = -\frac{\partial V_1}{\partial x_i} = -\frac{f}{\epsilon}\frac{\partial f}{\partial x_i} = -\lambda\frac{\partial f}{\partial x_i}. \tag{2.91}$$

This relation connects the linear and quadratic Lagrange multiplier methods: $\lambda = f/\epsilon$. In this sense the quadratic method handles small imagined violations of the $f = 0$ condition in the typical size of $f = \lambda\epsilon$. This consideration is sometimes called a "softening" of the constraint, resulting in a soft version of it. This becomes again hard in the $\epsilon \to 0$ limit. In fact in quantum field theory the quadratic term supports the naive Lagrange function, and ϵ is not set to zero from the beginning of the calculations.

2.7 Mapertuis Principle and Geodetic Motion

According to the original formulation of the Mapertuis principle the motion follows that path with minimal length. This was the principle of shortest path. Although for the general motion, under influence of forces, this proved to be erroneous, for the free motion it is valid. As such also can be applied to motions which are free in time intervals and only change at instants of zero measure. Its generalized form, valid for all motions in conservative force fields, was derived by Euler and Lagrange. This was a typical case, when an error helped to find the true relation! Please also note that this principle does not cite forces or energy, it states something about the orbits.

This principle applied to free motion trespassing borders of media leads to formulas akin to those used in geometrical optics. The main difference is that for the propagation of waves instead of mass points the principle of shortest time, the Fermat principle, applies.

For a free motion the Lagrange function contains a kinetic energy term only, the solution of the equation of motion ensures a velocity constant in time. Looking at the motion of a single mass point practically its kinetic energy time integral provides the action, it is to be minimized

$$\int T\,dt = \int \frac{mv^2}{2}\,dt = \text{minimum} \tag{2.92}$$

while the Mapertuis' shortest path principle reads as

$$\int m\,ds = \int mv\,dt = \text{minimum}. \tag{2.93}$$

For a constant velocity v, these two principles may agree, but for a general motion not.

For a motion under influence of forces the principle of shortest path is modified to a statement about a weighted length of that path. Essentially one seeks for a particular form of the action principle without any explicit reference to the time; only integrals over paths occur. During this the total energy is still conserved.

Let us consider a Lagrange function as follows:

$$L = \frac{1}{2} \sum_{i,k} a_{ik}(q)\dot{q}_i\dot{q}_k - V(q), \tag{2.94}$$

with q_i generalized coordinates and $V(q)$ potential energy being a function of the position only. The generalized momenta are

$$p_i = \frac{\partial L}{\partial \dot{q}_i} = \sum_k a_{ik}(q)\dot{q}_k. \tag{2.95}$$

The conserved energy based on Eq. (2.45) is given by

$$E = \frac{1}{2} \sum_{i,k} a_{ik}(q)\,\dot{q}_i\dot{q}_k + V(q). \tag{2.96}$$

This information can be used for rewriting the action $S = \int L\,dt$ in the new form:

$$S = \int \sum_{i,k} a_{ik}(q)\,\dot{q}_i \dot{q}_k \, dt - \int \left(\frac{1}{2} \sum_{i,k} a_{ik}(q)\,\dot{q}_i \dot{q}_k + V(q) \right) dt. \qquad (2.97)$$

Here in the first term the generalized momenta (2.95), in the second term the energy given by Eq. (2.96) is substituted in to get

$$S = \int \sum_i p_i \dot{q}_i \, dt - \int E \, dt. \qquad (2.98)$$

For constant energy, E, the second integral is proportional to the total time interval for the motion and is independent from the variations of the orbit. Therefore it suffices to vary the *reduced action*,

$$S' = S + E(t_2 - t_1), \qquad (2.99)$$

resulting in

$$\delta S' = \delta \left(\sum_i \int p_i dq_i \right) = 0. \qquad (2.100)$$

This is the Hamilton principle, an equivalent to the action principle for constant energy motion. The Eq. (2.100) does not refer to time coordinates or to derivatives against time. The varied quantities can be described by the variation of the paths followed during the motion in arbitrary parameterization. So this principle refers to paths and not to the time history of any motion.

In order to derive the formula called Mapertuis action to date we start with the expression

$$2\,(E - V(q)) = \sum_{i,k} a_{ik}(q)\,\dot{q}_i \dot{q}_k \qquad (2.101)$$

Integrating the right hand side (rhs) of this expression delivers the reduced action appearing in Eq. (2.100) as Hamilton's principle:

$$S' = \int \sum_i p_i \, dq_i = \int \sum_{i,k} a_{ik}(q)\,\dot{q}_i \dot{q}_k \, dt. \qquad (2.102)$$

Due to Eq. (2.101) this integrand can be viewed as the square root of the rhs and left hand side expression; by this trick we remove time derivatives and use coordinate differentials only:

$$S' = \int \sqrt{2\,(E - V(q)) \sum_{i,k} a_{ik}(q)\, dq_i dq_k}. \qquad (2.103)$$

We note that for a single mass point the square root of the weighted sum of coordinate differential square is by definition the differential arc length alongside the motion's path:

$$ds = \sqrt{\sum_{i,k} g_{ik}dq_idq_k},\tag{2.104}$$

with $a_{ik} = mg_{ik}$. In this case the reduced action really becomes the shortest path principle:

$$S' = \int \sqrt{2m\,(E - V(q))}\,ds = \text{extremum}\tag{2.105}$$

is the *Mapertuis principle*.

It is an interesting experience to investigate the equations for the orbit (motion's path) derived from the Mapertuis principle by functional variation. Omitting the constant factor $\sqrt{2m}$ for simplicity, the variation of the reduced Mapertuis action can be expressed by the orbit functions, $x_i(s)$, its variation and time derivatives:

$$\delta \int \sqrt{E - V}\,ds = -\sum_i \int \left(\frac{\partial V}{\partial x_i} \frac{\delta x_i}{2\sqrt{E - V}}\,ds - \sqrt{E - V}\,\frac{dx_i}{ds}\,d\delta x_i \right).\tag{2.106}$$

Here we have used the relation $ds^2 = \sum_i dx_i^2$ and its consequence $dsd\delta s = \sum_i dx_i\,d\delta x_i$, as well as the fundamental relation in variational calculus, according to which $\delta(ds) = d\delta s$. Integrating now the second term by parts we obtain the vanishing first variation condition as

$$2\sqrt{E - V}\,\frac{d}{ds}\left(\sqrt{E - V}\,\frac{dx_i}{ds}\right) = -\frac{\partial V}{\partial x_i} = F_i.\tag{2.107}$$

Here we denoted the force component, obtained as gradient of the potential energy, by F_i. This result has a wonderful geometrical interpretation. Observing that

$$2\sqrt{E - V}\,\frac{d}{ds}\sqrt{E - V} = -\frac{dV}{ds} = \sum_i F_i \frac{dx_i}{ds},\tag{2.108}$$

Equation (2.107) can be written as a relation with force components,

$$\sum_j F_j \frac{dx_j}{ds}\frac{dx_i}{ds} + 2(E - V)\frac{d^2x_i}{ds^2} = F_i.\tag{2.109}$$

The second derivative along the path replaces the Newtonian acceleration in this formula. Besides a term containing the square of first derivatives appears: This is analogous to the formula for *geodesic* motion in general relativity. An inertial force, pro-

Fig. 2.8 In a given point of a general $x_i(s)$ orbit the tangent vector, $t_i = dx_i/ds$, is normalized to unit length. Moreover its derivative is orthogonal to it and is connected to the radius of curvature in that point, $dt_i/ds = n_i/R$

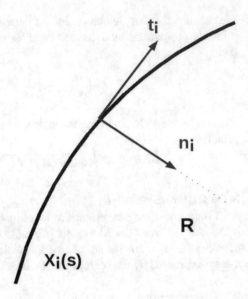

portional to the "velocity" square emerges as correction to direct forces. In this case the general Christoffel symbol is restricted to a special form, $\Gamma^k_{ij} = F_j \delta^k_i$ (Fig. 2.8).

Introducing the tangential vector to the path, $t_i = dx_i/ds$, it is easy to convince oneself that its length is unity: $\sum_i t_i t_i = 1$. Using this result the second derivative along the orbit can be expressed from Eq. (2.109):

$$\frac{d^2 x_i}{ds^2} = \frac{F_i - \left(\sum_j F_j t_j\right) t_i}{2(E - V)}. \tag{2.110}$$

This formula points out the followings: Only the orthogonal component of the force causes acceleration, $F_i^\perp = F_i - \left(\sum_j F_j t_j\right) t_i$. At the same time this second derivative is nothing else than the first derivative of the vector t_i tangent to the path. Due to the unit length of this vector its derivative is orthogonal to the tangent. Its direction can be denoted by another unit vector, n_i, and its magnitude as a curvature, $1/R$, of the path in a given point:

$$\frac{d^2 x_i}{ds^2} = \frac{dt_i}{ds} = \frac{1}{R} n_i. \tag{2.111}$$

Indeed R is the radius of a circle adjusted to the path in that point. Using this result and the expression $E - V = mv^2/2$ for the kinetic energy, Eq. (2.109) expresses the centripetal acceleration on a circle touching the path:

$$\frac{mv^2}{R} n_i = F_i^\perp. \tag{2.112}$$

The fall of an apple and the circular orbit of the moon are connected to each other; not only in a tale about Newton, but also by the Mapertuis principle—in its correct form due to Euler and Lagrange.

2.8 Legendre Transformation and Action—Angle Variables

We have mentioned the Legendre transformation earlier in connection with the relation between the Lagrange and Hamilton functions. During this transformation part of or all the descriptive variables will be replaced by new variables. The function governing the motion must be changed accordingly. By this special transformation the new variables are the old partial derivatives of the old governing function, and vice versa. Repeating a Legendre transformation towards the original variables we get back the original governing function.

Let us denote the "old" variables by u_i, the "new" ones by v_i. Given a function, $F(u_1, \ldots, u_n)$ the new variables are the $v_i = \frac{\partial F}{\partial u_i}$ partial derivatives, and the transformed function is given by

$$G = \sum_{i=1}^{n} u_i \frac{\partial F}{\partial u_i} - F. \tag{2.113}$$

Variations of functionals based on the new function contains these partial derivatives by construction: The total differential,

$$dG = \sum_i \frac{\partial G}{\partial v_i} v_i, \tag{2.114}$$

can be written in an equivalent form due to (2.113):

$$dG = \sum_i (u_i \, dv_i + du_i \, v_i) - dF = \sum_i \left[u_i \, dv_i + \left(v_i - \frac{\partial F}{\partial u_i} \right) du_i \right]. \tag{2.115}$$

It follows that the partial derivatives of the new function are exactly the old variables:

$$\frac{\partial G}{\partial v_i} = u_i. \tag{2.116}$$

The status of the "old" and "new" is hence symmetric.

The relation between the Lagrange and Hamilton functions in classical mechanics is special. Here $F = L$ and $G = H$, but only half of the variables are replaced: the role of generalized velocities, $u_i = \dot{q}_i$, will be overtaken by the generalized momenta, $v_i = p_i$. The opposite transformation produces a Hamilton function from a Lagrange function.

$$p_i = \frac{\partial L}{\partial \dot{q}_i}, \quad H = \sum_i p_i \dot{q}_i - L,$$

$$\dot{q}_i = \frac{\partial H}{\partial p_i}, \quad L = \sum_i p_i \dot{q}_i - H. \tag{2.117}$$

The action principle can be formulated based on the Hamiltonian, too,

$$S = \int_{t_1}^{t_2} \left(\sum_i p_i \dot{q}_i - H \right) dt = \sum_i p_i q_i \Big|_{t_1}^{t_2} - \int_{t_1}^{t_2} \left(\sum_i q_i \dot{p}_i + H \right) dt \tag{2.118}$$

Varying this action with respect to the $p_i(t)$ momenta and $q_i(t)$ position coordinates one obtains the equations of motion in the Hamiltonian dynamics:

$$\dot{q}_i - \frac{\partial H}{\partial p_i} = 0,$$

$$-\dot{p}_i - \frac{\partial H}{\partial q_i} = 0. \tag{2.119}$$

Regarding the (p_i, q_i) functions as coordinates in the phase space we obtain further interesting relations. From the equality of the mixed second partial derivatives of the Hamiltonian,

$$\frac{\partial}{\partial q_i} \frac{\partial}{\partial p_i} H = \frac{\partial}{\partial p_i} \frac{\partial}{\partial q_i} H, \tag{2.120}$$

it follows

$$\sum_i \left(\frac{\partial \dot{q}_i}{\partial q_i} + \frac{\partial \dot{p}_i}{\partial p_i} \right) = 0, \tag{2.121}$$

upon using the equations of motion (2.119). This can be interpreted as the divergenceless property of a current vector, $\mathbf{v} = (\dot{q}_i, \dot{p}_i)$, in $6N$-dimensional phase space for N particles:

$$\nabla \cdot \mathbf{v} = 0, \tag{2.122}$$

with $\nabla = (\frac{\partial}{\partial q_i}, \frac{\partial}{\partial p_i})$. The trajectories, describing the classical motion in phase space, constitute an incompressible (divergenceless) fluid. This is analogous to the theory of fluids, so the above result indicates the constancy of the phase space volume occupied by those moving points:

$$d\Omega = dq_1 \wedge dp_1 \ldots dq_n \wedge dp_n. \tag{2.123}$$

This is stated by the Liouville theorem. Its essence is that the solution of the equations of motion derived from an action principle for a short dt time step is always a *canonical transformation* leaving the phase space volume invariant.

A consequence of the Liouville theorem is Helmholtz's theorem: the number of vortices in the phase space liquid is constant, i.e. the reduced action calculated over closed orbits,

$$\Gamma = \oint \sum_i p_i \, dq_i, \qquad (2.124)$$

after reaching its extremum does not change in time. We have $d\Gamma/dt = 0$ along the solution curves of the equation of motion. Due to Stokes' theorem an integral over a closed curve can be calculated from the flux going through the circumvented area:

$$\oint \sum_i p_i \, dq_i = m \oint \mathbf{v} d\mathbf{s} = m \int (\nabla \times \mathbf{v}) d^2 \mathbf{f}. \qquad (2.125)$$

This flux integral provides the number of vortices, constant in time.

The above action calculated for periodic orbits, expressed in a (u, v) parameterization of the surface through which the flux is counted, leads to the construction of the Lagrange bracket:

$$\Gamma = \int \sum_i \left(\frac{\partial q_i}{\partial u} \frac{\partial p_i}{\partial v} - \frac{\partial p_i}{\partial u} \frac{\partial q_i}{\partial v} \right) du \wedge dv = \int [u, v] \, du \wedge dv. \qquad (2.126)$$

The inverse of the Lagrange bracket is the Poisson bracket, defined as

$$\{u, v\} := \sum_i \left(\frac{\partial u}{\partial q_i} \frac{\partial v}{\partial p_i} - \frac{\partial v}{\partial q_i} \frac{\partial u}{\partial p_i} \right). \qquad (2.127)$$

Based on these definitions it is easy to see that

$$\sum_a [u_i, v_a] \{u_k, v_a\} = \delta_{ik}. \qquad (2.128)$$

Finally we demonstrate that the actions calculated over periodic orbits themselves can be used as generalized momenta. Moreover, exactly this canonical description is the simplest, as it reflects the integrability of a motion.

We render a quantity measured in action units to each pair of generalized coordinate and momentum components:

$$J_k = \frac{1}{2\pi} \oint p_k dq_k. \qquad (2.129)$$

The total action is a sum of such contributions, $S = 2\pi \sum_k J_k$. This action can be regarded as a function of the general coordinates and the above J_k quantities: $S = S(q_1, \ldots, q_n, J_1, \ldots, J_n)$.

We associate partial derivatives to each J_k action parts

$$\vartheta_i = \frac{\partial S}{\partial J_i}. \tag{2.130}$$

These conjugate variables have no physical unit dimension, they are called angle variables. The integral of the angle variables for exactly one period of the motion is exactly 2π, as it is shown by the short derivation below:

$$\oint d\vartheta_i = \oint \sum_k \frac{\partial}{\partial J_i} \frac{\partial}{\partial q_k} S dq_k = \frac{\partial}{\partial J_i} \sum_k \oint p_k dq_k = 2\pi \sum_k \frac{\partial J_k}{\partial J_i} = 2\pi. \tag{2.131}$$

The goal is to find that very canonical transformation, which describes the motion instead of the original (q, p) phase space in terms of action–angle variables, (ϑ, J). Observing that the reduced action can be expressed as a function of the conserved energy, E,

$$S'(q, E) = \int p(q, E) dq, \tag{2.132}$$

the total differential, dS', can be written both in terms of old and new variables:

$$dS' = \frac{\partial S'}{\partial q} dq + \frac{\partial S'}{\partial p} dp = \frac{\partial S'}{\partial \vartheta} d\vartheta + \frac{\partial S'}{\partial J} dJ. \tag{2.133}$$

Now if one wishes to change from the variable $p = \frac{\partial S'}{\partial q}$ to J, then the conditions $\frac{\partial S'}{\partial p} = 0$ and $\frac{\partial S'}{\partial \vartheta} = 0$ have to be satisfied. From this the following Hamiltonian equations of motion are derived:

$$\dot{j} = 0, \qquad \dot{\vartheta} = \frac{\partial S'}{\partial J} = T \frac{\partial E}{\partial J}. \tag{2.134}$$

with T being the period time.

Summarizing, the J quantities are constant during the motion and the total change of the action is a sum of $2\pi J$-s. Therefore the conservative motions are characterized besides the conserved energy, E, also by the action components reminding to the angular momentum, J_k. In case the number of such J_k-s reaches the number of general coordinates, then the motion is integrable.

2.9 Fermat Principle

The laws of geometrical optics can be derived from the principle of *shortest time* proposed by Fermat; this is complementary to the Maupertuis principle. Here the most typical case is free propagation except at points of medium change. According to Fermat's principle

$$T = \int\limits_{t_1}^{t_2} dt = \int\limits_{x_1}^{x_2} \frac{dx}{v(x)} = \text{minimum}, \qquad (2.135)$$

with $v = c/n(x)$ being the—position dependent—speed of light in medium. For a constant rarefaction index, $n(x) = n$, the principles of shortest path and shortest time coincide.

For light propagation without any obstacle this principle ensures the shortest light ray path, in Euclidean geometry lines. This is the first law of geometrical optics. Reflection by a flat mirror is also easily treated by Fermat's principle. Positioning the final observer of the light ray in the origin of our coordinate system, $(0, 0)$, and a pointlike light source at $(0, h)$, then the $x = -d$ plane will be reached at the point $(-d, y)$. In this arrangement d can be varied. According to Fermat's principle the length of the light ray path is minimal:

$$cT = \sqrt{d^2 + (h - y)^2} + \sqrt{d^2 + y^2} = \text{minimum}. \qquad (2.136)$$

Derivating this expression with respect to y we obtain a condition for this extremum:

$$c\frac{\partial T}{\partial y} = \frac{y - h}{\sqrt{d^2 + (h - y)^2}} + \frac{y}{\sqrt{d^2 + y^2}} = 0. \qquad (2.137)$$

This result ensures the equality of incoming and reflected ray angles to the plane.

The Snellius–Descartes law of rarefaction on the border of two homogeneous but different media also can be derived from Fermat's principle. The above derivation

Fig. 2.9 The light ray breaking on the border of changing media satisfies the Fermat principle

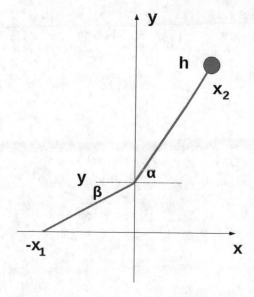

changes at a few points: now we follow light ray starting at (x_2, h) and arriving at $(-x_1, 0)$. A part of this path proceeds in a medium with optical index n, where the speed of light is c/n. The point to cross the border be at $(0, y)$. According to Fermat's principle

$$c\mathcal{T} = n\sqrt{x_1^2 + y^2} + \sqrt{x_2^2 + (h - y)^2} = \text{minimum.} \tag{2.138}$$

After derivation with respect to y, the quantity to be varied, we obtain

$$c\frac{\partial \mathcal{T}}{\partial y} = n\frac{y}{\sqrt{x_1^2 + y^2}} + \frac{y - h}{\sqrt{x_2^2 + (h - y)^2}} = n \sin \beta - \sin \alpha = 0. \tag{2.139}$$

From this the Snellius–Descartes law for the ratio of sines trivially follows (Fig. 2.9).

Finally we note that based on the similarity between the principles of shortest time and shortest path, it is suggestive to consider some correspondence between the laws of optics and mechanics: The Mapertuis principle is based on the integral $\int p\,dq$, the Fermat principle on $\int dq/v$. This suggests a correspondence like $p \sim 1/v$. Considering quantities of dimension length in both cases and the usual expression $p = mu$ for momenta, one arrives at the $uv = c^2$ reciprocity formula. This coincides with relation between group velocity and phase velocity known from wave propagation: an excellent motivation for de Broglie to state the particle wave duality. It is only an unexpected bonus on the top that due to this the quantization of action for periodic orbits in the atomic model by Bohr ensures that only an integer times the wavelength can be associated to an orbit of $2\pi r$ length.

Chapter 3
Gravity: The Optimal Curvature

Till date the theory of gravity went to the farthest on the way to interpret motion in terms of geometry: changing position does not happen in a separate time, but space and time follow a common, 4-dimensional geometry (special relativity) and the left alone bodies in this spacetime follow shortest paths, i.e. geodesics (general relativity). Intriguing a consequence of the principle stating the equivalence of descriptions of physical laws in either free falling or accelerating reference frames is that the local density of energy and momentum determines the geometry of spacetime, that the source of the curvature in spacetime is exactly the presence of the moving matter. The Einstein–Hilbert action minimizes the sum of covariant action integrals stemming from the curvature of spacetime and from the presence of matter or radiation energy.

3.1 Mapertuis Principle in Spacetime

As a first step we extend the Mapertuis principle related to the reduced action over paths to the four-dimensional spacetime; this will be the variational principle describing the (special) relativistic motion of a free mass point.

In this chapter, unless it may have a special meaning, we use the unit system $c = 1$, setting the lightspeed to one lightyear per year. After it has been revealed based on the Maxwell equations that the propagation speed of light waves in vacuum (without the presence of any material medium) is a natural constant, we may use it as a unity—following Albert Einstein. Adopting this convention the Mapertuis principle treats the lengths of paths in four-dimensional spacetime, i.e. that of worldlines, re-weighted in the free case only with the mass, as minimal—or at least as an integral to be varied.

© The Author(s), under exclusive license to Springer Nature Switzerland AG 2023
T. S. Biró, *Variational Principles in Physics*,
SpringerBriefs in Physics,
https://doi.org/10.1007/978-3-031-27876-1_3

The principle of special relativity, however, contains one more important notion: there is no absolute rest, for only the velocity of the relative motion counts in physics. This postulate, dating back to Galileo Galilei, together with the constancy of the lightspeed, enforces the Lorentz-transformation as the unique linear transformation between spacetime coordinates among moving reference frames. This transformation leaves a four-dimensional arc length invariant:

$$c^2 d\tau^2 = \|x\|^2 = c^2 dt^2 - dx^2 - dy^2 - dz^2, \tag{3.1}$$

assuming a so called flat spactime geometry with hyperbolic signature, also called a Minkowski spacetime. Minkowski himself—being a mathematician—suggested to view the fourth dimension pure imaginary, ict. We in this book refer to the length of the four-dimensional arc length, $cd\tau$ in (Eq. 3.1), as proper time (multiplied with $c = 1$). Since the ratio of space and time differentials deliver the actual velocity, $\mathbf{v} = \frac{d\mathbf{r}}{dt}$, one easily sees that

$$d\tau = dt\sqrt{1 - \mathbf{v}^2/c^2} = dt/\gamma, \tag{3.2}$$

or equivalently that a velocity-dependent factor γ, the Lorentz-factor appears between the proper time and coordinate-time: $\gamma = 1/\sqrt{1 - v^2/c^2}$. According to this $d\tau \leq dt$, so the apparent time difference is longer than the one measured by a comoving observer. This is the (special relativity) effect of time-dilation, among others this causes the observation of much more cosmic ray remnant muons near to Earth level as naively expected: although they decay in their proper time fast, looking from the Earth they have enough time to cross 80 km atmosphere. (If someone visits CERN, she/he should take the museum into program; there one can hear the detection of frequent muon hits.)

The relativistic action for a freely moving mass point m is

$$S = - \int mc^2 \, d\tau. \tag{3.3}$$

Viewing this as a coordinate-time integral we derive the Lagrange function for the free relativistic motion: starting from

$$S = \int L \, dt = - \int mc^2 \sqrt{1 - v^2/c^2} \, dt \tag{3.4}$$

one obtains

$$L = -mc^2\sqrt{1 - v^2/c^2} = -mc^2 + \frac{mv^2}{2} + \cdots \tag{3.5}$$

For low velocities ($v = |\mathbf{v}| \ll c$) the Lagrange function, disregarding the negative signed constant, gives back the familiar kinetic energy. The variation of the action (3.3) provides the equation of motion.

To obtain this we utilize the four-vector index notation for the coordinate differentials, $dx^i = (cdt, dx, dy, dz)$ for the covariant, while $dx_i = (cdt, -dx, -dy, -dz)$ for the contravariant version. The square of the proper time differential can now be expressed as

$$c^2 d\tau^2 = dx_i dx^i, \tag{3.6}$$

where using the Einstein convention lower and upper (contravariant and covariant) indices occurring twice are to be summed over for $i = 0, 1, 2, 3$. The variation of the proper time length, $cd\tau$, is therefore as follows

$$c\delta d\tau = \frac{dx_i \delta dx^i}{\sqrt{dx_j dx^j}} = \frac{dx_i}{cd\tau} \delta dx^i. \tag{3.7}$$

This form offers a plausible definition for the four-velocity,

$$u^i = \frac{dx_i}{d\tau}, \tag{3.8}$$

which by construction is the normalized length tangential four-vector to the world-line, $x^i(\tau)$, $u_i u^i = c^2$. Its components are $u^i = (\gamma c, \gamma \mathbf{v})$.

Using these notations the variation of the relativistic mass point action is given as

$$\delta S = -m \int u_i d\delta x^i = -\left. mu_i \delta x^i \right|_1^2 + m \int \delta x^i du_i. \tag{3.9}$$

Annulling the initial and final point contributions, occurring during the integration by parts, the equation of motion follows from this variation:

$$m \frac{du_i}{d\tau} = 0. \tag{3.10}$$

This is equivalent to the constancy of the velocity, $\mathbf{v}$.

On the other hand from the Lagrange function (3.5) the canonical momentum follows,

$$\mathbf{p} = \frac{\partial L}{\partial \mathbf{v}} = \frac{m\mathbf{v}}{\sqrt{1 - v^2/c^2}}. \tag{3.11}$$

By the help of this expression the Hamilton function, equal to the energy, is expressed:

$$H = E = \mathbf{p}\mathbf{v} - L = \frac{mc^2}{\sqrt{1 - v^2/c^2}}. \tag{3.12}$$

The momentum and the energy can be united in a single four-vector, too. This $p^i = (E/c, \mathbf{p}) = mu^i$ four momentum is constant during the free motion, so both energy and momentum is conserved. The length of this four-vector, $p_i p^i = m^2 c^2$, is a constant, characteristic to the mass point.

During a relativistic motion under the effect of forces the four-momentum will be changed. Analogous to the Newtonian formula for the force, $\mathbf{F} = \frac{d\mathbf{p}}{dt}$, one defines the following four-force:

$$g^i = \frac{dp^i}{d\tau} = m \frac{du^i}{d\tau}. \tag{3.13}$$

It is remarkable, that for a constant mass the four-force is orthogonal to the four-velocity:

$$u_i g^i = mu_i \frac{du^i}{d\tau} = \frac{m}{2} \frac{d}{d\tau}(u_i u^i) = 0. \tag{3.14}$$

Therefore the general form of a four-force,

$$g^i = (\gamma \, \mathbf{v} \, \mathbf{f}, \, \gamma \, c\mathbf{f}), \tag{3.15}$$

contains the power exerted by the same force, $\mathbf{f} \, \mathbf{v}$, too.

The Hamilton–Jacobi equations are based on the view of momenta as gradients of an action. In the relativistic case one uses its four-vector version,

$$p^i = -\frac{\partial S}{\partial x_i}, \tag{3.16}$$

while describing free motion in Eq. (3.10).

This free dispersion relation appears in the following Hamilton–Jacobi equation:

$$\frac{1}{c^2}\left(\frac{\partial S}{\partial t}\right)^2 - \left(\frac{\partial S}{\partial x}\right)^2 - \left(\frac{\partial S}{\partial y}\right)^2 - \left(\frac{\partial S}{\partial z}\right)^2 = m^2 c^2. \tag{3.17}$$

Subtracting the rest energy contribution from the relativistic energy, $E = -\frac{\partial S}{\partial t}$, one obtains a reduced action, $S' = S + mc^2 t$, satisfying the above Hamilton–Jacobi equation. In the nonrelativistic expansion, for $1/c^2 \to 0$ one obtains:

$$\frac{\partial S'}{\partial t} + \frac{1}{2m}\left(\frac{\partial S'}{\partial \mathbf{r}}\right)^2 = \frac{1}{2mc^2}\left(\frac{\partial S'}{\partial t}\right)^2. \tag{3.18}$$

Finally we emphasize that the action for a relativistically moving mass point is at the same time the four-dimensional generalization of the Hamilton action and expresses the principle of shortest time:

$$S = -\int p_i dx^i = \int \mathbf{p} d\mathbf{x} - \int E \, dt = -mc^2 \int d\tau. \tag{3.19}$$

3.2 Motion of Charges, Lorentz Force

The action for a point charge moving in electromagnetic field is easily derivable from that of the free mass point: only the four-momentum will be supplemented with a term, which couples the otherwise free system to the electromagnetic four-potential. The replacement $p_i \rightarrow p_i + eA_i$ results in the action

$$S = - \int (p_i + eA_i) \, dx^i; \tag{3.20}$$

Here the expression $(p_i + eA_i)$ used to be called "kinetic" momentum, too. In this section we use the $c = 1$ convention. Whenever quantum mechanical applications occur, where—as it will be shown later—the four-momentum is represented by the derivative operation with respect to the four-coordinate, $p_i = (\hbar/i)\partial_i$, this kinetic momentum corresponds to the gauge-covariant derivative:

$$p_i + eA_i = \frac{\hbar}{i} \left(\partial_i - \frac{e}{i\hbar} A_i \right). \tag{3.21}$$

The four-vector vector potential contains both the scalar and the vector potential parts of the classical electrodynamics, $A^i = (\Phi, \mathbf{A})$.

The first term in the action integral Eq. (3.20) can be rewritten in Mapertuis form, as it has been discussed in the previous section. By doing so we arrive at an action whose differential consists of products of Lorentz scalars as well as of Lorentz vectors:

$$S = - \int m \, d\tau - \int eA_i \, dx^i. \tag{3.22}$$

Here $eA_i \, dx^i$ (summed over the indices $i = 0, 1, 2, 3$ following the Einstein convention) and $md\tau = mu_i \, dx^i$ are differential forms of rank 1.

The Lagrange function can be read out form the formula (3.22):

$$L = -m\sqrt{1 - v^2} + e\mathbf{A}\mathbf{v} - e\Phi. \tag{3.23}$$

The derived Hamilton function,

$$H = \mathbf{v} \frac{\partial L}{\partial \mathbf{v}} - L = \frac{m}{\sqrt{1 - v^2}} + e \, \Phi \tag{3.24}$$

represents the energy of a point charge moving in an external electromagnetic field. It is obvious that relative to the energy of a neutral particle only a single term is added, proportional to the scalar potential. At the same time the canonical momentum derived from the Hamiltonian coincides with the kinetic momentum, and that does contain the vector potential:

$$\pi = \frac{\partial L}{\partial \mathbf{v}} = \frac{m\mathbf{v}}{\sqrt{1 - v^2}} + e\mathbf{A}. \tag{3.25}$$

The equation of motion is determined by applying the Euler–Lagrange formula:

$$\frac{d}{dt}\frac{\partial L}{\partial \mathbf{v}} - \frac{\partial L}{\partial \mathbf{r}} = 0. \tag{3.26}$$

In order to study this form, we transform the partial derivative of the Lagrangian with respect to the coordinate vector, $\mathbf{r}$, using a brief notation ∇ for this special operation:

$$\nabla L = e\nabla(\mathbf{v}\mathbf{A}) - e\nabla\Phi = e(\mathbf{v}\nabla)\mathbf{A} + e\mathbf{v} \times (\nabla \times \mathbf{A}) - e\nabla\Phi. \tag{3.27}$$

Using this result and the formula for the canonical momentum, (3.25), we ascertain that

$$\frac{d}{dt}(\mathbf{p} + e\mathbf{A}) = e(\mathbf{v}\nabla)\mathbf{A} + e\mathbf{v} \times (\nabla \times \mathbf{A}) - e\nabla\Phi. \tag{3.28}$$

Taking into account furthermore that the total time derivative of the vector potential, occurring on the left hand side of the above equation, can be written as a sum of changes due to the partial time derivative and a term due to the velocity, i.e.

$$\frac{d}{dt}\mathbf{A} = \frac{\partial}{\partial t}\mathbf{A} + (\mathbf{v}\nabla)\mathbf{A}, \tag{3.29}$$

in the final result some terms are cancelled:

$$\frac{d\mathbf{p}}{dt} = -e\frac{\partial \mathbf{A}}{\partial t} - e\nabla\Phi + e\mathbf{v} \times (\nabla \times \mathbf{A}). \tag{3.30}$$

Now, knowing expressions of the electric and magnetic field strengths in terms of potentials from electrodynamics, c.f.

$$\mathbf{E} = -\frac{\partial \mathbf{A}}{\partial t} - \nabla\Phi,$$
$$\mathbf{B} = \nabla \times \mathbf{A}, \tag{3.31}$$

finally we arrive at the expression of the Lorentz force:

$$\frac{d\mathbf{p}}{dt} = e\,(\mathbf{E} + \mathbf{v} \times \mathbf{B})\,. \tag{3.32}$$

The very same result can be achieved of course also without utilizing three-vector notation and field strength vectors. All we need is to start from the four-dimensional form of the action Eq. (3.20). The variation means the variation of the worldline orbits, $x^i(\tau)$. This has two effects: one directly through the coordinate differential featured in the action integral, and another, indirect one through the dependence of the vector potential components on the space and time coordinates:

$$\delta S = - \int (p_i + e A_i) d\delta x^i - e \int \frac{\partial A_j}{\partial x^i} \delta x^i \, dx^j = 0. \tag{3.33}$$

The first term we integrate by parts. The result is

$$\delta S = \int \left\{ d(p_i + e A_i) - e \frac{\partial A_j}{\partial x^i} dx^j \right\} \delta x^i = 0. \tag{3.34}$$

Expanding the differential of the four-vector potential in this formula we obtain the following condition for the vanishing of the action variation:

$$\frac{dp_i}{d\tau} + e \frac{\partial A_i}{\partial x^j} u^j - e \frac{\partial A_j}{\partial x^i} u^j = 0. \tag{3.35}$$

Introducing the four-dimensional field strength tensor,

$$F_{ij} = \frac{\partial A_j}{\partial x^i} - \frac{\partial A_i}{\partial x^j} = \partial_i A_j - \partial_j A_i, \tag{3.36}$$

being nothing else than the four-rotation of the four-vector potential, the following compact equation of motion for a point charge in the four-dimensional form emerges

$$\frac{dp_i}{d\tau} = e F_{ij} u^j. \tag{3.37}$$

Quoting the antisymmetric property of the field strength tensor, $F_{ji} = -F_{ij}$, based on the definition Eq. (3.36), one sees that the right hand side of the (3.37) delivers a genuine four-force, since $u^i(e F_{ij} u^j) = 0$ due to the antisymmetry.

3.3 The Weak (Newtonian) Gravitational Field

The potential energy of a mass m moving in gravitational field is strictly proportional to the same mass: $V(\mathbf{r}) = m\varphi(\mathbf{r})$. This very proportionality applies to inertial forces in accelerating systems and to the corresponding potential energy terms. This fact was summarized by Albert Einstein in the general theory of relativity: the laws of physics in any coordinate reference frame, accelerating relative to each other, describe the same content. This is the foundation of the theory of general relativity.

The gravitational force, due to its proportionality to the mass, cited above, counts as an inertial force (pseudo-force). The first experimental proof of this proportionality between the inertial and the gravitating mass was given by Roland Eotvos, and since then several modern experiments concluded the same with ever growing precision. This equivalence between the gravitating and inertial mass is the basis for all gravity theories. Not only for the—mathematically somewhat more complex—theory of general relativity, but also for the approximate, in weak gravitational field valid Newtonian theory.

The inertial mass, appearing in the kinetic energy expression in the Lagrange function, and the gravitating mass, included in the term for the potential energy, is treated as the same variable, m:

$$L = m \left(\frac{1}{2} \mathbf{v}^2 - \varphi(\mathbf{r}) \right). \tag{3.38}$$

Consequently, the common mass m cancels in the Euler–Lagrange equation of motion:

$$\frac{d}{dt} \mathbf{v} = -\nabla \varphi. \tag{3.39}$$

This means immediately that bodies with various m masses follow the same orbit in gravitational field! In particular the gravitational acceleration is independent of the mass: this was already asserted by Galilei with his experiments on slopes (letting objects fall from the tower in Pisa is perhaps purely a legend...).

As a simple example for an accelerating coordinate reference frame we investigate first a system rotating with constant ω angular velocity. What corrections arise in such a system? In a free system lacking forces, i.e. in an *inertial system* the relativistic geometry of spacetime follows the Minkowski geometry. Using $c = 1$ units,

$$d\tau^2 = dt^2 - dx^2 - dy^2 - dz^2. \tag{3.40}$$

Rotating the reference frame with constant ω angular velocity around the z axis we express the spacetime coordinates with those in the rotating system as follows:

$$x = x' \cos(\omega t') - y' \sin(\omega t'),$$
$$y = y' \cos(\omega t') + x' \sin(\omega t'),$$
$$z = z',$$
$$t = t'. \tag{3.41}$$

Based on this the coordinate differentials are obtained and the result is substituted into the Minkowski metric (3.40), delivering

$$d\tau^2 = \left[1 - \omega^2 (x'^2 + y'^2) \right] dt'^2 - dx'^2 - dy'^2 - dz'^2$$
$$+ 2\omega y' dx' dt' - 2\omega x' dy' dt'. \tag{3.42}$$

Here, in the coefficient of the term with dt'^2 besides the 1 the centrifugal "force" is recognized, while in the mixed (not pure quadratic) term reflects the effect of the Coriolis "force". The expression "force" is put between quotation marks because, as it is shown, these effects are only due to a time-dependent transformation of the spatial coordinates. They are in fact pseudo forces. In the four-dimensional spacetime this transformation is a point-dependent mixing of coordinates; a continuous and differentiable geometrical transformation. The source of this "force" is nothing else than the accelerating motion of the observer. Similarly gravity also does not have a special "cause". A free falling system is an inertial system, as it is experienced by the "weightless" motion of astronauts while orbiting around the Earth.

According to the general recipe the geometry of spacetime is modified compared to the gravitation free case. In a general metric the proper time differential length squared is given by

$$d\tau^2 = g_{ik}(x)\,dx^i dx^k, \tag{3.43}$$

The set of g_{ik} coefficients is called *metric tensor*. Based on the above construction this tensor is symmetric in all its indices. Whenever an experienced force effect (and its potential energy contribution, respectively) is proportional to the inertial mass, then this effect can be viewed in spacetime as a coordinate transformation: the effect is purely geometrical. The potential energy with de-factorized the m, is usually called a gravity *potential*, $\varphi(\mathbf{r})$.

A given metric tensor describes the geometry of spacetime, it contains all information with respect to apparent acceleration. In particular the coefficients, occurring in the expression for the arc length squared, i.e. the metric tensor, also serves as a starting point in calculating four-dimensional lengths, areas, volumes or four-volumes. In the so called *covariant* integrals the integration measure is given by $\sqrt{\det g}\, d^4x$ (due to the role that the Levi–Civitta symbol, $\epsilon_{ijk\ell}$ plays in calculating determinants). Finally, since there exists a smooth coordination of spacetime, where the components of the metric tensor are locally Minkowskian, i.e. only diagonal elements occur as $g_{ii} = (1, -1, -1, -1)$, the relation between general differentials and these specific differentials is given exactly by these metric tensor elements:

$$dx^i = \frac{\partial x^i}{\partial x'_k} dx'^k = g^{ik}(x')dx'^k. \tag{3.44}$$

The transformation rules of coordinate differentials are followed by further physical quantities. Vectors transforming this way are called *covariant*, but more complex quantities with several indices also can be constructed. These are baptized to *tensors*, as long as they transform in each of their indices "correctly", i.e. in analogy to coordinate differentials.

Conclusively, the general gravitational field is characterized by a general, space and time dependent metric tensor, $g_{ik}(x)$. In weak gravitational field the leading effect is included in the time-time-coordinate coefficient, g_{00}: this describes the *gravitational red shift*. The elapsed time at a fixed location ($dx^1 = dx^2 = dx^3 = 0$),

$$\mathcal{T} = \int d\tau = \int \sqrt{g_{00}}\,dx^0, \tag{3.45}$$

contains this component of the metric tensor. For motions with velocities much smaller than the lightspeed, and applying a weak gravitational potential, the Lagrange function is given by

$$L = -mc^2 + \frac{mv^2}{2} - m\varphi, \qquad (3.46)$$

and the corresponding action integral by

$$S = -mc \int \left(c - \frac{v^2}{2c} + \frac{\varphi}{c} \right) dt = -mc^2 \int d\tau. \qquad (3.47)$$

Here the second equality sign indicates that we interpret this action as that of a freely moving mass point in the language of proper coordinates. Hence $d\tau$ is expanded like

$$c\,d\tau = \left(c - \frac{v^2}{2c} + \frac{\varphi}{c} \right) dt. \qquad (3.48)$$

Using now that $\mathbf{v}dt = d\mathbf{r}$ and $v \ll c$, in leading order we get for the square of the proper time differential

$$c^2 d\tau^2 = (c^2 + 2\varphi)dt^2 - dx^2 - dy^2 - dz^2 + \mathcal{O}(1/c^2). \qquad (3.49)$$

Comparing this with the general expression (3.43) for the metric tensor, we realize that only the g_{00} component differs from the Minkowski spacetime values:

$$g_{00} = 1 + \frac{2\varphi}{c^2}. \qquad (3.50)$$

It is revealed that since the gravitational potential, φ, in the Newtonian gravity theory is a function of space, the spacetime is curved, while the three-dimensional Euclidean space itself is flat.

3.4   Geodesic Motion in General Gravitational Field

In this section we determine the equations for the free motion of a mass point, m, in case of having a general metric tensor. We trace back the variation of the Maupertuis action, $-m \int d\tau$, to the variation of the arc-length squared, which comprises information on the spacetime geometry. In the latter partial derivatives due to the space and time dependency of the metric tensor will play a role. Collecting them in a smart way one arrives at the equation describing the free motion as a *geodesic* acceleration. In this section we use units with $c = 1$.

The variation of the arc-length squared,

$$\delta d\tau^2 = 2d\tau \,\delta d\tau = \delta \left(g_{ik}dx^i dx^k\right) \tag{3.51}$$

receives several contributions when varying the $x^i(\tau)$ worldline coordinates; on the one hand due to the spacetime-dependence of the components of the metric tensor, $g_{ik}(x)$, on the other hand due to the change of coordinate differentials, dx^i and dx^k:

$$\delta \left(g_{ik}dx^i dx^k\right) = dx^i dx^k \frac{\partial g_{ik}}{\partial x^j}\delta x^j + 2g_{ik}dx^i \,\delta dx^k. \tag{3.52}$$

Using this relation the variation of the Mapertuis action becomes

$$\delta S = -m \int d\tau \left[\frac{1}{2}\frac{dx^i}{d\tau}\frac{dx^k}{d\tau}\frac{\partial g_{ik}}{\partial x^j}\delta x^j + g_{ik}\frac{dx^i}{d\tau}\frac{d\delta x^k}{d\tau}\right]. \tag{3.53}$$

Now we integrate the second term in this expression by parts. It results in

$$\delta S = -m \int d\tau \left[\frac{1}{2}\frac{dx^i}{d\tau}\frac{dx^k}{d\tau}\frac{\partial g_{ik}}{\partial x^j}\delta x^j - \frac{d}{d\tau}\left(g_{ik}\frac{dx^i}{d\tau}\right)\delta x^k\right]. \tag{3.54}$$

Finally we change the summation index from k to j in the second term, for making it possible to factorize the δx^j variation from the total expression under the integral. Introducing the four-velocity, $u^i = \frac{dx^i}{d\tau}$-t, we arrive at the following form of the Euler–Lagrange equation:

$$\frac{1}{2}u^i u^k \frac{\partial g_{ik}}{\partial x^j} - \frac{d}{d\tau}\left(g_{ij}u^i\right) = 0. \tag{3.55}$$

In order to reach the familiar form of the geodesic equation we need some more index magic. Applying the derivation with respect to τ to the product $g_{ij}u^j$, we apply the relation

$$\frac{dg_{ij}}{d\tau} = \frac{dx^k}{d\tau}\frac{\partial}{\partial x^k}g_{ij} = u^k \frac{\partial g_{ij}}{\partial x^k}. \tag{3.56}$$

Further we utilize the $i \leftrightarrow k$ index symmetry implied in the second term to obtain

$$u^i u^k \frac{\partial g_{ij}}{\partial x^k} = \frac{1}{2}u^i u^k \left(\frac{\partial g_{ij}}{\partial x^k} + \frac{\partial g_{kj}}{\partial x^i}\right). \tag{3.57}$$

Then we arrive at the geodesic motion describing accelerating motion due to the metric:

$$g_{ij}\frac{du^i}{d\tau} + \Gamma_{j,ik}u^i u^k = 0, \tag{3.58}$$

where the

$$\Gamma_{j,ik} = \frac{1}{2}\left(\frac{\partial g_{ij}}{\partial x^k} + \frac{\partial g_{kj}}{\partial x^i} - \frac{\partial g_{ik}}{\partial x^j}\right) \tag{3.59}$$

coefficients, the *Christoffel symbols*, can be determined from first derivatives of the metric tensor. These quantities do not construct a three index tensor, since their transformation properties do not follow those of co- or contravariant vectors.

In this equation the acceleration mixes with the square of velocity, similarly to the centrifugal acceleration terms discussed in the study of non-relativistic, curved orbits. The four-acceleration is caused by four-centrifugal forces.

3.5 Einstein-Hilbert Action

The Einstein–Hilbert action connects an optimum principle to the geometry of space-time: according to this the metric tensor is such that the corresponding spacetime possesses a minimal, covariant scalar curvature. Supplementing this principle with contributions from Lagrange functions describing further physical ingredients we obtain an action integral, from that the energy momentum tensor is connected to a tensor describing the local structure of spacetime. This is the Einstein equation.

In order to complete this program it pays to summarize a few elementary notions from differential geometry. So far we have considered arc length squares, which are one-dimensional objects. Their calculation from coordinate differentials is governed by the metric tensor, i.e. $ds^2 = g_{ik}dx^i dx^k$. From this we have derived the equation for the free motion aka the shortest worldline according to the Mapertuis principle.

Now we deal with questions about spacetime itself. Why is it so as it is? In this investigation we vary the metric tensor, and from the stationarity of the action integral against the variations in the g_{ik} components we determine the *field equations*. For finding the optimal smoothness we need the notion of curvature. And this, based on the component mixing in vectors and tensors while *parallel transported* in spacetime, leads through the operation of *covariant derivative* to the use of the *Riemann tensor*.

How can one judge whether a surface (or hypersurface or spacetime) is curved, without stepping out from it? That question was already answered by Gauss while studying the coordinates on curved two-dimensional surfaces. For an easier realization let us imagine a vector (a long pencil) which points into an external (embedding) space and moves there. Residents bound into the surface, however, can see only the shadow of it on the surface. The parallel transport, during which the origin of the vector moves on the curved surface, after completing a closed path does not bring this vector parallel to its initial stage. This experiment is easy to realize with the help of a big ball, or other curved surface like a saddle, and two pencils (Fig. 3.1).

The difference between the fixed and the transported vector after completing the closed loop will be proportional to (a) the length of the vector and (b) the path length. This is true for a surface with constant curvature, for other smooth curved spaces it

Fig. 3.1 Parallel trasnport of
vectors on a spherical
surface. Closing a path for
the transport the vector is not
parallel to its original, nor
the respective projections
onto the surface are

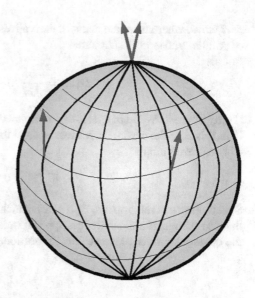

is true for an infinitesimally small path. The deviation in vector components can be
written as a linear combination of coordinate derivatives and the moved vector:

$$\Delta A_k = \cdots A_i dx^j, \tag{3.60}$$

where the ellipsis ... denotes unknown coefficients. Since in this problem three
directed vectors occur, these coefficients are quantities with three indices. These
happen to be the Christoffel–symbols, introduced in the geodesic equation:

$$\Delta A_k = \oint \Gamma^i_{kj} A_i dx^j. \tag{3.61}$$

However, not only the distance left behind but also the dimension of the piece of
the surface closed around can characterize the change in the transported vector. The
latter is a two-dimensional two-index quantity. A connection between these two
representations is made by the Stokes theorem. For a Δf_{jm} small surface element
we write

$$\Delta A_k = \frac{1}{2} \left[\frac{\partial (\Gamma^i_{km} A_i)}{\partial x^j} - \frac{\partial (\Gamma^i_{kj} A_i)}{\partial x^m} \right] \Delta f^{jm}. \tag{3.62}$$

Expanding the derivatives we obtain terms belonging to two classes: one containing
the derivatives of Christoffel-symbols and another one containing the derivatives of
the vector components. The factor $1/2$ is necessary due to the "double" counting of
the indices of the infinitesimal surface element.

The partial derivatives of the vector components express both physical changes and
location dependence. The "real", purely physical change is carried by the *covariant*

derivative, where from the result of derivatives with respect to coordinates the effect of parallel transport is subtracted:

$$D_j A_i = \frac{\partial}{\partial x^j} A_i - \Gamma_{ij}^n A_n. \tag{3.63}$$

If only a pure transport happens, the covariant derivative of the vector is zero, $D_j A_i = 0$. In this case the partial derivatives against the coordinates can be expressed by the Christoffel–symbols:

$$\frac{\partial A_i}{\partial x^j} = \Gamma_{ij}^n A_n. \tag{3.64}$$

Substituting this relation into Eq. (3.62), which reflects the Stokes theorem, we obtain that the change of the parallel transported vector is proportional to that vector and to the enclosed surface element. The proportionality coefficient has four indices:

$$\Delta A_k = \frac{1}{2} R_{k\ell m}^i A_i \Delta f^{\ell m}, \tag{3.65}$$

where

$$R_{k\ell m}^i = \frac{\partial \Gamma_{km}^i}{\partial x^\ell} - \frac{\partial \Gamma_{k\ell}^j}{\partial x^m} + \Gamma_{n\ell}^i \Gamma_{km}^n - \Gamma_{nm}^i \Gamma_{k\ell}^n. \tag{3.66}$$

This four index quantity is now a tensor, it transforms in each of its indices similar to coordinate differentials. A spacetime is called flat whenever all components of the Riemann tensor vanish, $R_{k\ell m}^i = 0$. Furthermore this tensor can be used for the derivation of the scalar curvature. In two dimensions this coincides with the Gauss curvature, the reciprocal of the product of radii in the tangential circles.

We flash another face of the Riemann tensor, related to gauge theory; i.e. we spell out the similarity of the mathematical arsenal above to the field strength tensor in electrodynamics. Namely, both can be written as the commutator of covariant derivative operations. Taking into account that

$$(DA_i)_\ell = \partial_\ell A_i - \Gamma_{i\ell}^k A_k = \left(\delta_i^k \partial_\ell - \Gamma_{i\ell}^k \right) A_k \tag{3.67}$$

the covariant derivative operation in the direction ℓ can be viewed as a two index matrix:

$$(D_\ell)_i^k = \delta_i^k \partial_\ell - \Gamma_{i\ell}^k. \tag{3.68}$$

This is a matrix in the indices ℓ and k, and at the same time an operator when acting on spacetime functions, like physical fields. Similarly, in electrodynamics the covariant four-derivative consists of a coordinate derivative and from a multiplication with a—non gauge invariant—vector potential:

$$D_\ell = \partial_\ell - ieA_\ell, \tag{3.69}$$

with $A_\ell = (\Phi, \mathbf{A})$ being the four vector potential. The correspondence in quantum mechanics between the momentum and the operation of derivation leads to the so called kinetic momentum,

$$\frac{\hbar}{i} D_\ell = P_\ell - e A_\ell, \tag{3.70}$$

which describes a coupling between charges and electromagnetic fields and modifies energy and momentum.

Since the covariant derivation is operator and matrix at the same time, its commutators are not trivial. Based on Eq. (3.66) one obtains

$$[D_k, D_\ell] A_i = R^m_{ik\ell} A_m. \tag{3.71}$$

It is noteworthy that here the indices k and ℓ belong to the directions of covariant derivations while the indices i and m denote the components of the vector, i.e. stand for the basis vectors denoting the axes directions. In the classical, four dimensional theory of gravity these indices stem from the same set, therefore here the distinction is only philosophical, in physical models inessential. Knowing all this the Riemann tensor can be written in a more abstract form, suppressing the basis direction indices:

$$R_{k\ell} = [D_k, D_\ell]. \tag{3.72}$$

This form is analogous to the connection of the electromagnetic field strength tensor with the four potential:

$$-ie F_{k\ell} = [D_k, D_\ell]. \tag{3.73}$$

Exactly in this sense $R_{k\ell}$ is the *gravitational field strength* tensor.

For setting a variational principle we are interested in the action integral. The starting point to obtain this is a scalar and generally covariant Lagrange density, whose space integral delivers the Lagrange function, and a further time-integral, the action. Moreover, using general spacetime coordinates the four-dimensional integration measure itself has to be made invariant; that is achieved by multiplying with the square root of the determinant of the metric tensor (since the metric tensor can be viewed as a Jacobian between two different coordination). The covariant integral measure is given as $\sqrt{-g} d^4 x$. Here g denotes the determinant of the g_{ik} metric tensor, the minus sign is due to the Minkowski metric with the diagonal signature $(+, -, -, -)$ to render the quantity under the square root positive.

The Einstein–Hilbert action is the covariant integral of the scalar curvature of spacetime. This scalar curvature can be obtained from the Riemann tensor by a double "index reduction", that means by summing over index pairs with the help of the metric tensor: $R_{ik} = R^\ell_{i\ell k}$ is the Ricci tensor, obtained by pairing one derivation direction index with a base axis index, while the scalar curvature, $R = R^i_i = g^{ik} R_{ki}$, by pairing the remaining base axis and derivative indices. The action describing gravitational field is a linear functional of this scalar curvature of spacetime:

$$S_g = K \int R \sqrt{-g} \, d^4x. \tag{3.74}$$

The coefficient K contains the lightspeed as a natural constant and $K = -c^3/(16\pi\kappa)$. Although this action functional is linear in the scalar curvature it is highly nonlinear in terms of the metric tensor. This is the origin of several difficulties.

In describing the "matter" part the central element is a Lagrange density, $\mathcal{L}$, composed from contributions stemming from physical fields, e.g. electromagnetic fields or other fields describing elementary particles, sometimes classical fluids. These contributions to the action are also integrated over the spacetime using the same covariant measure:

$$S_m = \int \mathcal{L} \sqrt{-g} \, d^4x. \tag{3.75}$$

Gravitating systems are described by the common use of gravity field and matter terms, in fact the Hilbert–Einstein action proper is $S = S_g + S_m$. As it was mentioned, the severe nonlinearity in the determinant of the metric tensor couples the "material" Lagrange density to the spacetime geometry. The consequence is not only a modification of equations known from other areas of physics in gravitational field, but also the fact that the source of the gravity is just in these material fields. The coupling between matter and spacetime via the metric tensor makes this theory tensorial, characterized by exactly ten independent quantity due to the symmetry inherent in the metric tensor, $g_{ik} = g_{ki}$. The source of gravity can be comprised into another symmetric tensor, the energy-momentum tensor, $T_{ik} = T_{ki}$.

To derive the energy-momentum tensor we regard the action term, S_m. Varying this against the metric tensor gives

$$\delta S_m = \int \left[\frac{\partial \mathcal{L}\sqrt{-g}}{\partial g^{ik}} - \frac{\partial}{\partial x^\ell} \frac{\partial \mathcal{L}\sqrt{-g}}{\partial \frac{\partial g^{ik}}{\partial x^\ell}} \right] \delta g^{ik} \, d^4x. \tag{3.76}$$

Here the components of the metric tensor, g^{ik}, play the role of the general coordinates to be varied. The corresponding general velocities are its derivatives with respect to coordinates (not only the time derivatives because an integration both in time and space is made for obtaining the action). Therefore the above result is already in a form prepared for the Euler–Lagrange field equation, since the δg^{ik} has been factorized. Still, the integration must be rewritten in a general covariant form with the measure $\sqrt{-g} d^4x$. Beyond that "all what is needed" is the determination of the derivative of the determinant of the metric tensor with respect to its tensor components.

In this endeavour it can help that the mixed upper and lower indexed metric tensor is always unity, therefore its inverse coincides with its lifted component form: $g_{ik} = (g^{-1})^{ik}$. This is simply the abstract identity, $g = gg^{-1}g$. Taking this into account and observing that a determinant can be expanded according to corresponding sub-determinants, who are all another term in the inverse matrix up to a determinant factor, we gain the following relation for the variation of the square root of the determinant:

$$\delta\sqrt{-g} = -\frac{1}{2\sqrt{-g}}\delta g = -\frac{1}{2}\sqrt{-g}\,g_{ik}\delta g^{ik}. \tag{3.77}$$

Collecting all these we arrive at the following variation of the matter action part:

$$\delta S_m = \frac{1}{2}\int T_{ik}\delta g^{ik}\sqrt{-g}\,d^4x, \tag{3.78}$$

where the energy-momentum tensor is given by

$$T_{ik} = \frac{1}{\sqrt{-g}}\frac{\partial \mathcal{L}\sqrt{-g}}{\partial g^{ik}} - \frac{1}{\sqrt{-g}}\frac{\partial}{\partial x^\ell}\frac{\partial \mathcal{L}\sqrt{-g}}{\partial\frac{\partial g^{ik}}{\partial x^\ell}}. \tag{3.79}$$

This definition results from the variation of the general covariant action integral with respect to the components of the metric tensor. But does this result in agreement with other derived results for the energy and momentum utilizing Noether's theorem? First of all is the energy and momentum described by T_{ik} conserved?

It can be shown that this tensor satisfies a local, covariant conservation law. Let's start from the variation of the matter action, (3.78). We interpret the components of the g^{ik} metric tensor as coefficients in a linear coordinate transformation between locally geodetic ξ^k and the general x^i coordinates,

$$g^{ik} = \frac{\partial \xi^i}{\partial x_k}. \tag{3.80}$$

In a locally geodetic coordinate system all Christoffel-symbol values vanish, (their derivatives cannot vanish at the same time), therefore in a given point the covariant and the simple partial derivatives coincide. Taking into account furthermore the index symmetry one asserts that

$$\delta g^{ik} = \frac{1}{2}\left(D^k\delta\xi^i + D^i\delta\xi^k\right) \tag{3.81}$$

is the variation of the metric tensor. The rules of integration by parts are valid for covariant integrals with the measure $\sqrt{-g}d^4x$, with the only difference that the simple partial derivatives have to be replaced by covariant derivatives. With this knowledge the variation of the matter action part can be transformed into

$$\delta S_m = -\int D_k(T_i^k)\delta\xi^i\sqrt{-g}\,d^4x. \tag{3.82}$$

Consequently the T_{ik} energy-momentum tensor satisfies the following local, covariant conservation law

$$D_k T^{ki} = 0. \tag{3.83}$$

For obtaining the full theory of gravity one needs to vary the action term containing
the scalar curvature, too. This we trace back to the variation of the Ricci and the
metric tensor. Here the former contains second derivatives of the latter. The variation

$$\delta S_g / K = \delta \int (R\sqrt{-g})\, d^4x = \delta \int g^{ik} R_{ik} \sqrt{-g}\, d^4x. \tag{3.84}$$

contains three types of terms: the direct variation of the metric tensor, the variation
of the Ricci tensor, and the variation of the square root of the determinant. For the
last term we utilize the result (3.77). We obtain

$$\delta S_g / K = \int \left(R_{ik} - \frac{1}{2} g_{ik} R \right) \delta g^{ik} \sqrt{-g}\, d^4x + \int g^{ik} \delta R_{ik} \sqrt{-g}\, d^4x. \tag{3.85}$$

Terms collected in the bracket define the famous Einstein tensor:

$$G_{ik} = R_{ik} - \frac{1}{2} g_{ik} R. \tag{3.86}$$

One needs to prove that the contribution of the separated term is zero. Again we pro-
vide this proof in a locally geodetic coordinate system, where the Christoffel-symbols
vanish and the covariant derivative coincides with the simple partial derivative. The
varied term contains only the derivatives of the Christoffel-symbols

$$g^{ik} \delta R_{ik} = g^{ik} \left(\frac{\partial}{\partial x^\ell} \delta \Gamma^\ell_{ik} - \frac{\partial}{\partial x^k} \delta \Gamma^\ell_{i\ell} \right). \tag{3.87}$$

In the last term above an exchange of the summation indices ℓ and k does not change
the result. Therefore

$$g^{ik} \delta R_{ik} = g^{ik} \frac{\partial}{\partial x^\ell} \delta \Gamma^\ell_{ik} - g^{i\ell} \frac{\partial}{\partial x^\ell} \delta \Gamma^k_{ik}. \tag{3.88}$$

Let us inspect now the four vector $w^\ell = g^{ik} \delta \Gamma^\ell_{ik} - g^{i\ell} \delta \Gamma^k_{ik}$. In a locally geodesic
coordinate system it is true that

$$g^{ik} \delta R_{ik} = \frac{\partial}{\partial x^\ell} w^\ell, \tag{3.89}$$

so the investigated variation is a total four divergence. Since w^ℓ is indeed a vector in
a general coordinate system this variation is covariant four divergence:

$$g^{ik} \delta R_{ik} = D_\ell w^\ell = \frac{1}{\sqrt{-g}} \frac{\partial}{\partial x^\ell} (w^\ell \sqrt{-g}). \tag{3.90}$$

The covariant four-dimensional integral of this covariant four divergence is a hyper-surface term, it can be put to zero on a far enough integration boundary. Finally we have obtained the variation of the total Einstein–Hilbert action as

$$\delta S_g + \delta S_m = -K \int \left(G_{ik} - \frac{2}{K} T_{ik} \right) \delta g^{ik} \sqrt{-g} d^4 x = 0. \tag{3.91}$$

From this form it immediately follows the famous Einstein equation

$$G_{ik} = R_{ik} - \frac{1}{2} g_{ik} R = \frac{8 \pi \kappa}{c^4} T_{ik}. \tag{3.92}$$

Chapter 4
Electrodynamics: Forces, Fields, Waves

It is not customary in university courses to derive electrodynamics from a variational principle. Nevertheless two of the Maxwell equations, those with a source term, and the light and radio wave propagation equations can be obtained from a single variational principle. Historically the development of electrodynamics went on another way. However, in order to help the deepening in the applications of variational principles, in this book we derive the basic laws of electricity, magnetism and their relation from those.

Our starting point is the minimization of the global energy hidden in physical fields. This will be supported by various conditions, which however may vary point by point since they contain the fields varying from a point to another. One of the most characteristic consequences is that the corresponding Lagrange multiplicators are continuously many, instead of a simple index they are functions of space position vectors, $\mathbf{r}$, and time, t. The variational principle, an energy to be minimal in the static case or an action to be extreme in the dynamical case, sheds light to new relations between important phenomena, which are not at all obvious on the level of field equations constructed from experiments step by step. Among others the famous displacement current, introduced by Maxwell, will be connected to the law of induction.

After the discussion of static electric and magnetic fields, respectively, we shall construct the action for electrodynamics and derive the Maxwell equations from it. Following this we discover a duality between electric and magnetic fields, with a corresponding arsenal of algebraic notations. Finally the propagation of electromagnetic waves will be derived in this chapter from a variant of the action which is supported by a term owing to the gauge fixing in the Lorenz gauge. We close the review of electrodynamics by spelling out the connection from the Maxwell equations to the spin one.

© The Author(s), under exclusive license to Springer Nature Switzerland AG 2023
T. S. Biró, *Variational Principles in Physics*,
SpringerBriefs in Physics,
https://doi.org/10.1007/978-3-031-27876-1_4

4.1 Electrostatic Gauss Principle

It is a typical problem in electrostatics to calculate electric fields corresponding to a given ensemble of point charges or to a continuous charge density. The pertinent variational principle here is the minimum of the energy carried by the static electric fields. One considers this energy as a volume integral of the electric energy density,

$$H = \int \frac{\epsilon_0}{2} \mathbf{E}^2 \, d^3r = \text{minimum.} \tag{4.1}$$

Here ϵ_0 is the dielectric constant of the vacuum. Of course, since the integrand is everywhere positive or zero, the minimal value of the above integral is also zero. In the presence of charges, however, with a general electric charge density ρ, the electric fields must satisfy the Gauss law:

$$\nabla \mathbf{E} = \rho/\epsilon_0. \tag{4.2}$$

Introducing the Lagrange multiplicator field, $\lambda = \epsilon_0 \Phi(\mathbf{r})$, associated to the condition prescribed by Gauss law, the full variational principle in electrostatics becomes

$$W = \epsilon_0 \int \left[\frac{1}{2}\mathbf{E}^2 - \Phi(\nabla \mathbf{E} - \rho/\epsilon_0) \right] d^3r = \text{minimum.} \tag{4.3}$$

The functional derivative of this quantity with respect to the electric field delivers a new equation:

$$\frac{\delta W}{\delta \mathbf{E}} = \epsilon_0 \, (\mathbf{E} + \nabla \Phi) = 0. \tag{4.4}$$

The variation against the Lagrange multiplicator, Φ, trivially leads back to the Gauss law.

According to the variational solution, the electric field strength is the negative gradient of a scalar field, $\mathbf{E} = -\nabla\Phi$. This scalar field is the electric potential. An immediate consequence is the basic equation of electrostatics

$$\nabla \times \mathbf{E} = 0. \tag{4.5}$$

It is interesting to express the electrostatic energy, W, in terms of the Lagrange multiplicator field in stead of the original field strength tensor. Substituting the result (4.4) into the integral (4.3) we obtain

$$W_0 = W|_{\frac{\delta W}{\delta \mathbf{E}}=0} = \epsilon_0 \int \left[\frac{1}{2}(\nabla\Phi)^2 + \Phi\nabla^2\Phi + \rho\Phi/\epsilon_0 \right] d^3r. \tag{4.6}$$

Here the ∇^2 denotes a double application of the ∇ operation, summing up contributions from each vector index. Its short name is Laplace operator. Now the first term in Eq. (4.6) can be integrated by parts:

$$W_0 = \int \left[\epsilon_0 \left(\frac{1}{2} \Phi \nabla^2 \Phi \right) + \rho \Phi \right] d^3 r. \tag{4.7}$$

Using furthermore the Poisson equation, derived from the $\mathbf{E} = -\nabla \Phi$ relation and the Gauss law,

$$\nabla^2 \Phi = -\rho/\epsilon_0, \tag{4.8}$$

the final result is given as

$$W_0 = \frac{1}{2} \int \rho \Phi \, d^3 r. \tag{4.9}$$

The total energy put into the electric field, H, based on the Eq. (4.8) also coincides with this integrated value:

$$H_0 = \frac{\epsilon_0}{2} \int (\nabla \Phi)^2 \, d^3 r = \frac{1}{2} \int \rho \Phi \, d^3 r, \tag{4.10}$$

after integration by parts and using the Poisson equation.

It is noteworthy that—not using the Poisson equation—Eq. (4.6) can be integrated by parts also in the second term instead of the first. Then the quantity W_0 to be varied is expressed as another functional, that of Φ:

$$W_0 = \epsilon_0 \int \left[-\frac{1}{2} (\nabla \Phi)^2 + \rho \Phi/\epsilon_0 \right] d^3 r. \tag{4.11}$$

Varying this functional with respect to Φ, of course, gives back the Poisson equation (4.8).

4.2 Magnetostatics

It is a natural question whether something similar can be formulated for the connection between stationary currents and magnetic fields. The $\mathbf{B}$ magnetic field[1] carries an energy analogous to the case in electrostatics:

$$H = \int \frac{1}{2} \mathbf{B}^2 \, d^3 r. \tag{4.12}$$

---

[1] Not distinguishing here magnetic induction and field, treating the magnetic permeability as unity.

Here is also a condition to be added with Lagrange multiplier, connecting the currents with derivatives of the field strength. This is formulated in Ampére's law,

$$\nabla \times \mathbf{B} - \mathbf{j} = 0, \tag{4.13}$$

with $\mathbf{j}$ being the current density vector. This extra condition is a vector (with three components in this case), therefore the Lagrange multiplier is also a vector field:

$$W = \int \left[\frac{1}{2} \mathbf{B}^2 - \mathbf{A} \left(\nabla \times \mathbf{B} - \mathbf{j} \right) \right] d^3r = \text{minimum} \tag{4.14}$$

is the variational principle of magnetostatics. The variable field $\mathbf{A}(\mathbf{r}, t)$ is recognized as the usual vector potential. Namely the functional derivative of W defined in Eq. (4.14) delivers the correspondence

$$\frac{\delta W}{\delta \mathbf{B}} = \mathbf{B} - \nabla \times \mathbf{A} = 0. \tag{4.15}$$

In order to arrive at this formula with the correct signs one has to keep in mind that not only the integration by parts results in a relative minus, but also the permutation of indices in the Levi-Civitta symbol inherent in the definition of rotation:

$$\int \mathbf{A} \left(\nabla \times \delta \mathbf{B} \right) d^3r = \int \varepsilon_{ijk} A^i \nabla^j (\delta B^k) \, d^3r$$

$$= - \int (\varepsilon_{ijk} \nabla^j A^i) \delta B^k \, d^3r = \int (\varepsilon_{ijk} \nabla^i A^j) \delta B^k \, d^3r. \tag{4.16}$$

Viewing from variational principles, both the scalar potential introduced in the electrostatics and the vector potential in magnetostatics play the role of Lagrange multipliers.

Similarly to the electrostatics the $\mathbf{B}$ field can be eliminated by substituting Eq. (4.15) into Eq. (4.13):

$$\nabla \times (\nabla \times \mathbf{A}) = \mathbf{j}. \tag{4.17}$$

The rotation of a rotation can be expressed as the gradient of the divergence minus the Laplace operator:

$$\nabla \times (\nabla \times \mathbf{A}) = \nabla(\nabla \mathbf{A}) - \Delta \mathbf{A}, \tag{4.18}$$

with the Laplace operator $\Delta = \nabla^2$. Meanwhile due to the relation $\mathbf{B} = \nabla \times \mathbf{A}$ the vector potential can be increased by a pure gradient of a scalar function without changing our equations:

$$\mathbf{B} = \nabla \times \mathbf{A} = \nabla \times (\mathbf{A} + \nabla \chi), \tag{4.19}$$

since the rotation of a gradient is identically zero. This so called *gauge freedom* is which authorizes us to choose an appropriate vector potential to our calculations, that simplifies our formula. If we are able to reach that the divergence of the new vector potential, $\mathbf{A}' = \mathbf{A} + \nabla\chi$, vanishes, i.e. $\nabla\mathbf{A}' = 0$, then we obtain a purely Laplacian equation for each component:

$$\Delta\mathbf{A}' = \mathbf{j}. \tag{4.20}$$

This condition can be reached if the harmonic condition for the function χ,

$$\nabla(\mathbf{A} + \nabla\chi) = 0 \tag{4.21}$$

is satisfied. This is equivalent to

$$\Delta\chi = -\nabla\mathbf{A}. \tag{4.22}$$

As one easily recognizes a little freedom of choice is left over, but it is restricted to solutions of the homogeneous Laplace equation.

4.3 Variational Principle of Electrodynamics

It was James Clark Maxwell's discovery that besides the current density occurring in Eq. (4.13) there is another current, tagged as displacement current, proportional to the velocity of change of the electric field in time. Using reduced units, in which dielectric constants and magnetic permeability are unity, we simply write

$$\nabla \times \mathbf{B} = \mathbf{j} + \dot{\mathbf{E}} \tag{4.23}$$

Here the overdot denotes a *partial* derivative against time. The task is to correct the current density, $\mathbf{j}$, in the above discussed variational principles. However, since electric and magnetic effects combine with each other in the dynamics, we have to construct a single dynamical variational principle.

Following experimental results not the sum but the difference of electric and magnetic field energies become parts of this new principle, analogous to having the difference of kinetic and potential energy terms in the Lagrange function. The variational principle of the full electrodynamics is therefore constructed as the action in mechanics:

$$S = \int dt\, \{W_e - W_m\} \tag{4.24}$$

with

$$W_e = \int d^3r \left[\frac{1}{2} \mathbf{E}^2 - \Phi \left(\nabla \mathbf{E} - \rho \right) \right],$$

$$W_m = \int d^3r \left[\frac{1}{2} \mathbf{B}^2 - \mathbf{A} \left(\nabla \times \mathbf{B} - \mathbf{j} - \dot{\mathbf{E}} \right) \right]. \tag{4.25}$$

In these formulas we use such units where the dielectric constant and the magnetic permeability are unity, this means measuring charge and current in appropriate units. This renders the lightspeed also to unity, as in studies of the theory of relativity it is frequently done.

The action integral of electrodynamics is over space and time, the action functional is invariant. This relativistic invariance against the necessary Lorentz transformations, however, stay hidden in the present notations. At the same time it is to be noted already at this point that the terms connected to external sources, i.e. charge and current densities, are collected in a form identical to the scalar product of Lorentz four vectors: $(\rho, \mathbf{j}) \cdot (\Phi, \mathbf{A})$.

The full electrodynamics action (4.24) is constructed by involving the Gauss and Ampére laws. Nothing else has to be added, the description of the induction discovered by Faraday automatically follows from this action by functional derivation. Variations against Φ and $\mathbf{A}$ give back the conditions, but variations against the field strength vectors $\mathbf{E}$ and $\mathbf{B}$ lead to new relations:

$$\frac{\delta S}{\delta \mathbf{E}} = \mathbf{E} + \nabla \Phi + \dot{\mathbf{A}} = 0,$$

$$\frac{\delta S}{\delta \mathbf{B}} = -\mathbf{B} + \nabla \times \mathbf{A} = 0. \tag{4.26}$$

All this can be obtained by doing integration by parts both in spatial and temporal integrations. From the two lines in Eq. (4.26) it follows the Faraday equation describing the magnetic induction: only the rotation of the first line has to be compared with the (partial) time derivative of the second line:

$$\nabla \times \mathbf{E} = -\nabla \times \dot{\mathbf{A}} = -\dot{\mathbf{B}}. \tag{4.27}$$

Another consequence, already valid in magnetostatics, is the vanishing divergence of the magnetic induction vector field:

$$\nabla \mathbf{B} = 0. \tag{4.28}$$

This special Maxwell equation points towards the impossibility of magnetic monopoles (magnets alike point charges), since otherwise there would be a source term in the right hand side of the above equation, analog to the equation for the electric field. Phenomena known in nature so far do not hint to the existence of magnetic monopoles. At the same time it is worth to think of them theoretically, because then the symmetry between electric and magnetic quantities would be complete. Such assumptions may discover new, unexpected relations.

4.4 Electric—Magnetic Duality

Even if no magnetic monopole is available in our laboratories, theoretically we can double the world of electric and magnetic phenomena: imagine that all Lagrange multipliers discussed so far has a magnetic counterpart, $(\Phi_m, \mathbf{A}_m)$, and also the charges and currents, $(\rho_m, \mathbf{j}_m)$. Yet, in the universe we experience their value is zero due to some unseen cause.

The symmetry explored and completed in this way is known under the name electric—magnetic *duality*. In the absence of electric charges and currents, so in the majority of space in our universe, this symmetry is realized. Applying this symmetry the action contains the field energy integrals and both electric and magnetic secondary conditions:

$$S_d = \int dt \int d^3r \left(\frac{1}{2} \mathbf{E}^2 - F_e - F_m - \frac{1}{2} \mathbf{B}^2 + G_e + G_m \right), \qquad (4.29)$$

with

$$
\begin{aligned}
F_e &= \Phi \left(\nabla \mathbf{E} - \rho \right), \\
F_m &= \Phi_m \left(\nabla \mathbf{B} + \rho_m \right), \\
G_e &= \mathbf{A} \left(\nabla \times \mathbf{B} - \mathbf{j} - \dot{\mathbf{E}} \right), \\
G_m &= \mathbf{A}_m \left(\nabla \times \mathbf{E} + \mathbf{j}_m + \dot{\mathbf{B}} \right).
\end{aligned}
\qquad (4.30)
$$

The list above contains, besides the original electric contributions, indexed by e, also a dual counterpart, indexed by m. In this setting the variations against Lagrange multipliers directly deliver the original Maxwell equations in the language of field strengths, at least when we substitute $\rho_m = 0$ and $\mathbf{j}_m = 0$.

$$
\begin{aligned}
\frac{\delta S_d}{\delta \Phi} &= -\nabla \mathbf{E} + \rho = 0, \\
\frac{\delta S_d}{\delta \Phi_m} &= -\nabla \mathbf{B} - \rho_m = 0, \\
\frac{\delta S_d}{\delta \mathbf{A}} &= \nabla \times \mathbf{B} - \mathbf{j} - \dot{\mathbf{E}} = 0, \\
\frac{\delta S_d}{\delta \mathbf{A}_m} &= \nabla \times \mathbf{E} + \mathbf{j}_m + \dot{\mathbf{B}} = 0.
\end{aligned}
\qquad (4.31)
$$

In this interpretation the variations against the field strength components deliver relations between potentials (Lagrange multipliers) and field strengths:

$$
\begin{aligned}
\frac{\delta S_d}{\delta \mathbf{E}} &= \mathbf{E} + \nabla \Phi + \dot{\mathbf{A}} + \nabla \times \mathbf{A_m} = 0, \\
\frac{\delta S_d}{\delta \mathbf{B}} &= -\mathbf{B} + \nabla \times \mathbf{A} - \dot{\mathbf{A}}_m + \nabla \Phi_m = 0.
\end{aligned}
\qquad (4.32)
$$

Studying the Maxwell equations without source terms reveals that a so called *dual transformation* exchanges the magnetic and electric fields in an antisymmetric manner. The following transformation,

$$\mathbf{E}' = -\mathbf{B}, \qquad \mathbf{B}' = \mathbf{E}, \tag{4.33}$$

changes the action into its negative counterpart. This means that the extremizing variational equations, reflecting the observable dynamics, do not change.

When using all (real and imagined) source terms we need more: in order to bring S_d defined in (4.29) into its negative also the following changes have to apply:

$$\rho' = \rho_m, \quad \rho'_m = -\rho, \qquad \mathbf{j}' = -\mathbf{j}_m, \quad \mathbf{j}'_m = \mathbf{j},$$
$$\Phi' = \Phi_m, \quad \Phi'_m = -\Phi, \qquad \mathbf{A}' = -\mathbf{A}_m, \quad \mathbf{A}'_m = \mathbf{A}. \tag{4.34}$$

It is extraordinarily simple and beautiful to take into account this duality by using a complex field strength vector,

$$\mathbf{F} = \mathbf{E} + \mathrm{i}\mathbf{B}, \tag{4.35}$$

with i being the imaginary unit. Doing so the dual transformation is simply a multiplication with i: $\mathbf{F}' = \mathrm{i}\mathbf{F}$. The complete dual action itself is easier to be handled when using complex quantities. For the sake of brevity we mean from now on the complex construction by the unindexed quantities,

$$\mathbf{A} = \mathbf{A}_e + \mathrm{i}\mathbf{A}_m, \qquad \Phi = \Phi_e - \mathrm{i}\Phi_m,$$
$$\mathbf{j} = \mathbf{j}_e + \mathrm{i}\mathbf{j}_m, \qquad \rho = \rho_e - \mathrm{i}\rho_m. \tag{4.36}$$

In this case the dual transformations listed under Eq. (4.34) are also achieved by a simple multiplication with i.

Let us now formulate a new action principle, without the magnetic charges and their current but using the complex field strength:

$$S'_d = \int dt d^3r \left[\frac{1}{2}\mathbf{F}^2 - \Phi \left(\nabla \mathbf{F} - \rho \right) - \mathbf{A} \left(\mathrm{i}\nabla \times \mathbf{F} + \dot{\mathbf{F}} + \mathbf{j} \right) \right], \tag{4.37}$$

and its extremum delivers electrodynamics. In the above formula conditions resembling the Gauss and Ampére laws are involved in complex notation. The square of the complex field strength,

$$\mathbf{F}^2 = \mathbf{E}^2 - \mathbf{B}^2 + 2\mathrm{i}\mathbf{E} \cdot \mathbf{B} \tag{4.38}$$

gives back the traditional term $E^2 - B^2$ as its real part, but its imaginary part is a new contribution at this level. We have to convince ourselves that this new term would not spoil the original theory!

From here on we annul magnetic monopole like terms, indexed with m, but keep the complex notation. Now the variations against complex functions lead to the following results:

Variation against the complex field strength

$$\frac{\delta S_d'}{\delta \mathbf{F}} = \mathbf{F} + \nabla \Phi + \dot{\mathbf{A}} - i\nabla \times \mathbf{A} = 0. \qquad (4.39)$$

The real part of this complex equation connects the electric, its imaginary part the magnetic fields with the Lagrange multiplicators: $\mathbf{E} + \nabla \Phi + \dot{\mathbf{A}} = 0$ and $\mathbf{B} - \nabla \times \mathbf{A} = 0$.
Variation against the scalar potential results in

$$\frac{\delta S_d'}{\delta \Phi} = -\nabla \mathbf{F} + \rho = 0, \qquad (4.40)$$

whose real part is the Gauss law, $\nabla \mathbf{E} = \rho$, and its imaginary part expresses the absence of magnetic monopoles, $\nabla \mathbf{B} = 0$.
Finally variation against the vector potential delivers,

$$\frac{\delta S_d'}{\delta \mathbf{A}} = -i\nabla \times \mathbf{F} - \dot{\mathbf{F}} - \mathbf{j} = 0, \qquad (4.41)$$

whose real part is the Ampére-Maxwell equation, $\nabla \times \mathbf{B} = \dot{\mathbf{E}} + \mathbf{j}$ while its imaginary part is the Faraday equation, $\nabla \times \mathbf{E} = -\dot{\mathbf{B}}$.

The real part of the complex action (4.37) coincides with the action of electrodynamics (4.24). What about its imaginary part? Is

$$\Im m \, S_d' = \int dt d^3 r \left[\mathbf{E} \cdot \mathbf{B} - \Phi \, \nabla \mathbf{B} - \mathbf{A} \left(\nabla \times \mathbf{E} + \dot{\mathbf{B}} \right) \right]. \qquad (4.42)$$

disturbing? To our luck the variation of the imaginary part of the action does not lead to new equations:

$$\frac{\delta}{\delta \mathbf{B}} \Im m \, S_d' = \mathbf{E} + \nabla \Phi + \dot{\mathbf{A}} = \frac{\delta}{\delta \mathbf{E}} \Re e \, S_d',$$
$$\frac{\delta}{\delta \mathbf{E}} \Im m \, S_d' = \mathbf{B} - \nabla \times \mathbf{A} = -\frac{\delta}{\delta \mathbf{B}} \Re e \, S_d', \qquad (4.43)$$

The structure of the above functional derivative relations is a full analogy to the Cauchy–Riemann relations known from the theory of complex analytic functions. Summarizing briefly, the complex action is a complex analytic functional of the complex field strength.

4.5 Electromagnetic Waves

The Maxwell equations not only unified the descriptions of electric and magnetic behavior, but they predicted a brand new physical phenomenon: the electromagnetic waves. A specially narrow wavelength range of it is the visible light. The existence of waves must be recognizable on the level of the variational principle, too. The clue is the Poisson equation; there we have seen that formulating the action in terms of the Lagrange multiplier fields (in electrostatics the scalar potential, Φ) leads to a second order partial differential equation. A similar form can be achieved also for the magnetic sector, but there we exploited the gauge freedom, too. In this section we rearrange the action of the full electrodynamics containing the complex field strength in a way that the description of electromagnetic waves will become obvious.

In the action functional presented in Eq. (4.37) we integrate by parts those terms which contain the complex field strength, $\mathbf{F}$. There are three such terms: that containing a divergence, a rotation and a partial time derivative.

$$S'_d = \int dt d^3r \left[\frac{1}{2}\mathbf{F}^2 - \Phi(\underbrace{\nabla \mathbf{F}}_{parc.int.} - \rho) - \mathbf{A}(i \underbrace{\nabla \times \mathbf{F} + \dot{\mathbf{F}}}_{parc.int.} + \mathbf{j}) \right] \qquad (4.44)$$

As a result we obtain a combination of terms linear and quadratic in the complex field strength,

$$S'_d = \int dt d^3r \left[\frac{1}{2}\mathbf{F}^2 + \mathbf{F} \cdot \nabla\Phi + \mathbf{F} \cdot \dot{\mathbf{A}} - i\mathbf{F} \cdot \nabla \times \mathbf{A} + (\rho\Phi - \mathbf{j} \cdot \mathbf{A}) \right], \qquad (4.45)$$

that can be transformed to a complete square:

$$S'_d = \int dt d^3r \left[\frac{1}{2}(\mathbf{F} + \mathbf{\Xi})^2 - \frac{1}{2}\mathbf{\Xi}^2 + (\rho\Phi - \mathbf{j} \cdot \mathbf{A}) \right], \qquad (4.46)$$

with

$$\mathbf{\Xi} = \nabla\Phi + \dot{\mathbf{A}} - i\nabla \times \mathbf{A}. \qquad (4.47)$$

It is obvious that the vanishing of the first quadratic term at $\mathbf{F} = -\mathbf{\Xi}$ delivers the known relations between field strengths and potentials. This at the same time prepares the optimization with a complete square Lagrange density: the vanishing of the variation against the complex field strength, $\mathbf{F}$. The remaining action, similarly to the tricks at the derivation of the Poisson equation, gets simplified:

$$S_0 = \int dt d^3r \left[\rho\,\Phi - \mathbf{j} \cdot \mathbf{A} - \frac{1}{2}\mathbf{\Xi}^2 \right]. \qquad (4.48)$$

Using the expression (4.47) for the complex potential, Ξ, the real and imaginary part of the above action reads as:

$$\Re S_0 = \int dt d^3r \left[\rho\,\Phi - \mathbf{j}\cdot\mathbf{A} - \frac{1}{2}\left(\nabla\Phi + \dot{\mathbf{A}}\right)^2 + \frac{1}{2}\left(\nabla\times\mathbf{A}\right)^2 \right],$$

$$\Im S_0 = \int dt d^3r \left[\left(\nabla\Phi + \dot{\mathbf{A}}\right)\cdot\left(\nabla\times\mathbf{A}\right) \right]. \tag{4.49}$$

For our analysis of the variations, we start with the imaginary part:

$$\frac{\delta}{\delta\Phi}\,\Im S_0 = -\nabla(\nabla\times\mathbf{A}) = 0,$$

$$\frac{\delta}{\delta\mathbf{A}}\,\Im S_0 = \nabla\times\left(\dot{\mathbf{A}}\right) - (\nabla\times\mathbf{A})^{\cdot} = \mathbf{0}. \tag{4.50}$$

We conclude that the variations both against the scalar and against the vector potential lead to identities! Therefore the dynamics described by the complex reduced action or its real part only are identical to each other. This complex action is equivalent to its real part.

The equations responsible for the dynamics of the potentials are therefore derived from the real part of the complex action. We have to arrive at the same result if we vary the full complex action against the scalar and vector potential and then take the real part. The variations are

$$\frac{\delta}{\delta\Phi}S_0 = \nabla\left(\nabla\Phi + \dot{\mathbf{A}} - i\nabla\times\mathbf{A}\right) + \rho = 0,$$

$$\frac{\delta}{\delta\mathbf{A}}S_0 = \frac{\partial}{\partial t}\left(\nabla\Phi + \dot{\mathbf{A}} - i\nabla\times\mathbf{A}\right)$$
$$+ i\nabla\times\left(\nabla\Phi + \dot{\mathbf{A}} - i\nabla\times\mathbf{A}\right) - \mathbf{j} = 0. \tag{4.51}$$

The real parts of the above equations contain the second time and spatial derivatives of the potentials:

$$\Delta\,\Phi + \nabla\dot{\mathbf{A}} = -\rho,$$
$$\nabla\dot{\Phi} + \ddot{\mathbf{A}} + \nabla\times(\nabla\times\mathbf{A}) = \mathbf{j}. \tag{4.52}$$

Here in the second equation the rotation of the rotation of the vector potential appears, which can be expressed using the Laplace operator. So we obtain the following two equations with charge and current densities as source terms:

$$-\left(\dot{\Phi} + \nabla\mathbf{A}\right)^{\cdot} + \ddot{\Phi} - \Delta\,\Phi = \rho,$$
$$\nabla\left(\dot{\Phi} + \nabla\mathbf{A}\right) + \ddot{\mathbf{A}} - \Delta\,A = \mathbf{j}. \tag{4.53}$$

In order to recognize wave propagation in these equations we have to use a further auxiliary condition for the divergence of the vector potential—analogous to the derivation of the magnetostatic Poisson equation. This gauge freedom allows us to apply the Lorenz gauge fixing, the so called "radiation gauge":

$$\nabla \mathbf{A} + \dot{\Phi} = 0. \tag{4.54}$$

Using this a magic happens: in the first and second line of Eq. (4.53) the first terms vanish and the rest become to pure wave equations for the scalar and vector operator:

$$\ddot{\Phi} - \Delta \Phi = \rho,$$
$$\ddot{\mathbf{A}} - \Delta \mathbf{A} = \mathbf{j}. \tag{4.55}$$

Particular solutions to these wave equations in absence of the sources, i.e. charges and currents, are the electromagnetic waves (EM waves). They can be observed experimentally at numerous frequencies and are utilized in various technologies. Beyond the visible light also the radio and microwaves and the infrared radiation are parts of modern technology, penetrating into the life of civilized humanity. Our picture about the far universe on the other hand singularly is from the detection of these waves until a few years ago gravitational waves have also been detected. These EM waves propagate in vacuum with the speed of light, taken as unity in this chapter. They are describing vibrations transverse to the direction of propagation. At the same time they carry energy and momentum, exactly in the direction of their propagation. The relation between energy and momentum, according to the frequency—wave number vector relation, the dispersion relation of the EM wave, belongs to a massless particle, the *photon*.

4.6   Variational Principle with Gauge Fixing

In the previous section we have experienced that the EM wave equations could be derived from the complex variational principle in electrodynamics only if a particular gauge fixing, the Lorenz gauge, is applied. This gauge fixing choice, however, is not only a condition on the level of equations, but it can be built in and should be built in into the action itself. Almost all variational action can be supplemented by side conditions by using the method of Lagrange. The Lorenz gauge fixing we include in a quadratic form, with a special coefficient, $1/2$, now:

$$\Re S_0' = \int dt\, d^3 r \left[-\frac{1}{2} \left(\nabla \Phi + \dot{\mathbf{A}} \right)^2 + \frac{1}{2} \left(\dot{\Phi} + \nabla \mathbf{A} \right)^2 + \frac{1}{2} \left(\nabla \times \mathbf{A} \right)^2 \right. $$
$$\left. + \; (\rho\, \Phi - \mathbf{j} \mathbf{A}) \right]. \tag{4.56}$$

In this case the sum of the first two terms can be intelligently rearranged so that the mixed terms together constitute a total time derivative. Such parts of the Lagrangians can be omitted from the action. With more detail,

$$
-\frac{1}{2}\left(\nabla\Phi + \dot{\mathbf{A}}\right)^2 + \frac{1}{2}\left(\dot{\Phi} + \nabla\mathbf{A}\right)^2 =
$$
$$
\frac{1}{2}\left(\dot{\Phi}^2 - (\nabla\Phi)^2\right) - \frac{1}{2}\left(\dot{\mathbf{A}}^2 - (\nabla\mathbf{A})^2\right) + \left(\dot{\Phi}\nabla\mathbf{A} - \dot{\mathbf{A}}\nabla\Phi\right), \qquad (4.57)
$$

from which after integrating by parts the contribution of the last bracket, once in the time once in the space integral, we obtain:

$$
\int dt\, d^3r \left(\dot{\Phi}\nabla\mathbf{A} - \dot{\mathbf{A}}\nabla\Phi\right) = \int dt\, d^3r \left(\dot{\Phi}\nabla\mathbf{A} + \Phi\nabla\dot{\mathbf{A}}\right). \qquad (4.58)
$$

This expression contains the time derivative of a product, whose contribution after integration again can be left out from the action:

$$
\int dt\, d^3r \left(\dot{\Phi}\nabla\mathbf{A} + \Phi\nabla\dot{\mathbf{A}}\right) = \int dt \frac{\partial}{\partial t} \int d^3r (\Phi\nabla\mathbf{A}) = 0. \qquad (4.59)
$$

What is left from the action of electrodynamics by including the Lorenz gauge in quadratic form is as follows:

$$
\Re S_0' = \int dt\, d^3r \left[\frac{1}{2}\left(\dot{\Phi}^2 - (\nabla\Phi)^2\right) - \frac{1}{2}\left(\dot{\mathbf{A}}^2 - (\nabla\mathbf{A})^2 - (\nabla\times\mathbf{A})^2\right)\right.
$$
$$
\left. + \; (\rho\,\Phi - \mathbf{j}\,\mathbf{A})\,\right]. \qquad (4.60)
$$

This formula is now more and more symmetric with respect to the scalar and vector potential, we start to see Lorentz-invariant structures. The time-derivative square minus space derivative squares type combinations as well as the last term reminding to the product of four vectors reflect the signature of the metric in the Minkowski spacetime.

The gauge fixed reduced action, Eq. (4.60), varied against the potentials now directly deliver wave equations:

$$
\frac{\delta}{\delta\Phi}\,\Re S_0' = -\ddot{\Phi} + \Delta\,\Phi + \rho = 0,
$$
$$
\frac{\delta}{\delta\mathbf{A}}\,\Re S_0' = \ddot{\mathbf{A}} - \Delta A - \mathbf{j} = 0. \qquad (4.61)
$$

4.7   Spin One

The complex field strength notation promotes another interesting observation beyond the above discussed dual symmetry and Lorentz transformation behavior: The electromagnetic field equations in vacuum in complex notation can look as a quantum mechanical setup. In the Heisenberg representation of energy and momentum the complex field strength behaves as a polarization vector part of a wave function in the Schrödinger picture. This means that the electromagnetic action and therefore the Maxwell equations induce a spin one wave. The impossibility of the longitudinal photon is an extra condition on the top of that; the sourceless Gauss equation enforces it.

The Maxwell equations in vacuum for the complex field strength, $\mathbf{F} = \mathbf{E} + i\mathbf{B}$, are as follows:

$$\dot{\mathbf{F}} + i(\nabla \times \mathbf{F}) = 0, \qquad \nabla \mathbf{F} = 0. \tag{4.62}$$

Writing the rotation in components and multiplying by $i\hbar$ we obtain

$$\left(\delta_{ik} i\hbar \frac{\partial}{\partial t} - \hbar\, \epsilon_{ijk} \nabla^j \right) F^k = 0. \tag{4.63}$$

This equation has the same form as a time dependent Schrödinger equation. Albert Einstein "almost" created quantum theory by studying the photons, what he could not make a peace with was the *uncertainty relation*. The classical theory describing the photons is so similar to the Schrödinger formalism, only the operator interpretation appears strained. In the next chapter, dealing with the variational principle behind quantum mechanics, we shall see that neither the operator algebra nor the uncertainty relation is needed for obtaining the Schrödinger equation.

Introducing now the momentum operator component-wise, $p^j = \frac{\hbar}{i}\nabla^j$, Eq. (4.63) delivers a particular Hamilton operator:

$$H_{ik} = i\, \epsilon_{ijk} p^j. \tag{4.64}$$

This operator in matrix representation has a spectrum; the eigenvalues can be calculated by solving the characteristic equation

$$\det \left| i\, \epsilon_{ijk} p^j - \omega\, \delta_{ik} \right| = 0. \tag{4.65}$$

Written component by component it reads as

$$D = \det \begin{vmatrix} -\omega & ip_3 & -ip_2 \\ -ip_3 & -\omega & ip_1 \\ ip_2 & -ip_1 & -\omega \end{vmatrix} = 0. \tag{4.66}$$

Expanding the determinant by its rows we obtain the following result:

$$D = -\omega\,(\omega^2 - p_1^2) - \mathrm{i}p_3(\mathrm{i}p_3\omega + p_1 p_2) + (-\mathrm{i}p_2)(-p_3 p_1 + \mathrm{i}p_2\omega). \qquad (4.67)$$

In this expression the triple product $\mathrm{i}p_1 p_2 p_3$ occurs twice, but with opposite signs, these terms cancel. In the rest the sum of the squares of the p_i components occurs, this is the length square of the momentum vector. So the characteristic equation simplifies to

$$D = -\omega\,(\omega^2 - |\mathbf{p}|^2) = 0. \qquad (4.68)$$

We have three different energy eigenvalues as solution: i) $\omega = +|\mathbf{p}|$, ii) $\omega = -|\mathbf{p}|$ and iii) $\omega = 0$. Taking into account the two scalar Maxwell equations, $\nabla \mathbf{F} = 0$, too, it is revealed that a given combination of the eigenvectors must not occur in the solution: the equation $p_k F_k = 0$ indicates the impossibility of a longitudinally polarized photon. A solution with this property exists only for the eigenvalues $\omega = \pm|\mathbf{p}|$, for the $\omega = 0$ eigenvalue no such solution can be found with a normalized eigenvector.

The latter is easy to prove using the vectorial form,

$$\mathrm{i}\mathbf{p} \times \mathbf{F} = \omega \mathbf{F}, \qquad \mathrm{i}\mathbf{p} \cdot \mathbf{F} = 0, \qquad (4.69)$$

in case of $\omega = 0$ would mean that $\mathbf{F}$ both parallel and orthogonal to $\mathbf{p}$. Therefore $\mathbf{F} = 0$ classically, describing no field just vacuum. Such a null vector cannot be normalized to one, too.

The absence of a longitudinally polarized degree of freedom is true for all massless particles, described by waves propagating with the speed of light in vacuum. Such waves can have only two polarization states, independent from the number of vector components in the field equations. The two physical eigenvalues discussed above belong to left and right handed helicities. The left- and right-handedness is mathematically described by the eigenvalue sign of the Levi-Civitta symbol, ϵ_{ijk}. Which is positive, is a question of agreement.

4.8 Quaternion Formalism

We have experienced in the previous section how to use complex fields in exploring the symmetries of the electrodynamical action. The question arises now, whether further algebraic structures may also hide in the system of Maxwell equations, or in the action in the background, respectively. The answer is positive as we know from the history of special relativity, since the formulas of the Lorentz transformation, mixing the components of four vectors linearly, are inherent in the equations of electrodynamics.

On the level of the action our task is to find that elegant and concise structure whose composed scalar serves as the Lagrange density. The theory of special relativity, unifying time and space coordinates and potential components into four vectors, hints at this. Such structures had occurred, however, well before the theory of relativity and the Minkowski geometry in physics and mathematics. Especially as results of mathematical aspirations to generalize the so successful complex numbers. Hamilton called these four element structures *quaternions* while Gauss tagged them as *hypercomplex* numbers. These two structures are equivalent. All operations with such quantities can be derived from the corresponding operations of certain "unit elements", spanning a ring over the algebra.

In general complex numbers can be given by two real numbers, and their multiplication rules are based on the corresponding rules for the imaginary unit, i, and the real unit, 1. The product of general complex numbers, $(x, y) = x + iy$, owing to the rule $i^2 = -1$ follows the following pattern:

$$(x_1 + iy_1)(x_2 + iy_2) = (x_1x_2 - y_1y_2) + i(x_1y_2 + y_1x_2). \tag{4.70}$$

Following the thoughts of Gauss, a hypercomplex number consists of a "real" and an "imaginary" part, both parts being complex numbers themselves. But this should happen by using another imaginary unit. Let us denote it by j, and the hypercomplex numbers by $p + jq$. Here $p = x + iy$ and $q = u + iv$. So we obtain four different kind of terms in the hypercomplex quantity, where the product of the two imaginary units, $k = ji$ also occurs:

$$p + jq = (x + iy) + j(u + iv) = x + iy + ju + kv. \tag{4.71}$$

Beyond the real number components we have three, pairwise "orthogonal" imaginary units, i, j, k. Using them as "axes" we obtain the quaternion representation.

However, one problem is still left. If $k = ji$ is another imaginary unit, then what is ij ? In a product of two different hypercomplex numbers namely both orders of i and j occur. We have to find a rational agreement, which closes the quaternion algebra multiplication table. A proof can be given that the correct choice is only $ij = -k$, if we want to conserve the imaginary unit property of the square being minus one for all, $i^2 = j^2 = k^2 = -1$. Since according to the above rules we have $ijk = ijji = -i^2 = 1$, therefore $-ij = ijk^2 = ijk \cdot k = k$ so $ij = -k$.

From the same requirement it follows that exchanging the order of different imaginary units a sign change enters, i.e. these quaternion imaginary units *anticommute*. Counting them with the index notation of Minkowski geometry from 0 to 3, it is purposeful to define $e_0 = 1, e_1 = j, e_2 = i$ and $e_3 = k$, for containing the ring of the quaternion algebra as

$$e_0^2 = e_0, \qquad e_0e_i = e_ie_0 = e_i,$$
$$e_ie_j = -\delta_{ij}e_0 + \epsilon_{ijk}e_k, \tag{4.72}$$

with $i = 1, 2, 3$ imaginary (space) directions.

So a general quaternion can be viewed as a four real number structure, but also as a combined structure of a scalar and a three-vector:

$$Q = q_0 e_0 + q_i e_i = (q_0, \mathbf{q}). \tag{4.73}$$

We shall see that the scalar–vector notation fits to spacetime and to the electromagnetic equations well, the symmetries of the action are pictured in the multiplication rule above. The product of two general quaternions, based on the above rule, becomes

$$Q\,P = (q_0 p_0 - q_i p_i) e_0 + (q_0 p_k + q_k p_0 + \epsilon_{ijk} q_i p_j) e_k, \tag{4.74}$$

or in 3-vector notation

$$(q_0, \mathbf{q}) \cdot (p_0, \mathbf{p}) = (q_0 p_0 - \mathbf{q} \cdot \mathbf{p}, \; q_0 \mathbf{p} + \mathbf{q} p_0 + \mathbf{q} \times \mathbf{p}). \tag{4.75}$$

This rules combine the scalar and vector product applied to three-vectors. That shows that they are fit for exploring the algebraic structure of the Maxwell equations.

The multiplication rules (4.72) can be represented in various ways. The most frugal one uses the Pauli matrices with a factor i and the 2×2 unit matrix:

$$e_0 = \begin{pmatrix} 1 & 0 \\ 0 & 1 \end{pmatrix} \qquad e_1 = \begin{pmatrix} i & 0 \\ 0 & -i \end{pmatrix}$$

$$e_2 = \begin{pmatrix} 0 & 1 \\ -1 & 0 \end{pmatrix} \qquad e_3 = \begin{pmatrix} 0 & i \\ i & 0 \end{pmatrix} \tag{4.76}$$

The quaternion conjugate changes the sign of the vector part (all imaginary parts), $Q^\dagger = (q_0, -\mathbf{q})$-t, so it can be used for creating the length square as a sure scalar quantity:

$$\|Q\|^2 = Q Q^\dagger = Q^\dagger Q = (q_0^2 + \mathbf{q}^2, 0) \tag{4.77}$$

is symmetric, contains only a real component and it is then and only then zero if all components are zero. On the other hand the square of a quaternion, like the square of a complex number, is still a general quaternion, $Q^2 = (q_0^2 - \mathbf{q}^2, 2 q_0 \mathbf{q})$.

The quaternions inherent in the electrodynamics, however, contain their vectorial part multiplied with another imaginary unit, i, due to the Minkowski metric. This i is different from all e_k quaternion units! Only the vector component is pure imaginary. The four vector potential therefore is represented by a quaternion, $A = (\Phi, i\mathbf{A})$, the derivations against space and time coordinates by the partial derivative quaternion, $\partial = (\partial_t, i\nabla)$. In this section the field strength is also a quaternion, for some later purpose we define it as a conjugate one:

$$F = \partial^\dagger A^\dagger. \tag{4.78}$$

Applying now the quaternion multiplication rule we obtain

$$F = (\partial_t, -i\nabla) \cdot (\Phi, -i\mathbf{A}) = (\dot{\Phi} + \nabla\mathbf{A}, -i\dot{\mathbf{A}} - i\nabla\Phi - \nabla \times \mathbf{A}). \tag{4.79}$$

Examining the scalar and vector components of the result, $F = (G, i\mathbf{F})$, we can determine that the 3-vector part is the complex field strength, $\mathbf{F} = \mathbf{E} + i\mathbf{B}$, and the scalar part, $G = \dot{\Phi} + \nabla\mathbf{A}$, is exactly the quantity set to zero in the Lorenz gauge.

The half of the length square of the field strength quaternion delivers the Lagrange density part without the coupling to charges and currents:

$$\frac{1}{2}F^\dagger F = \frac{1}{2}\left(G^2 - \mathbf{F}^2\right), \tag{4.80}$$

that substituting the expressions for G and F from Eq. (4.56) gives

$$\frac{1}{2}F^\dagger F = \frac{1}{2}\left(\dot{\Phi} + \nabla\mathbf{A}\right)^2 - \frac{1}{2}\left(\dot{\mathbf{A}} + \nabla\Phi - i\nabla \times \mathbf{A}\right)^2. \tag{4.81}$$

For the coupling to charges and currents we utilize the four current quaternion, $J = (\rho, i\mathbf{j})$. This, multiplied with the four vector potential quaternion, still contains some vectorial components. We cure this problem by adding the conjugate term.

$$JA^\dagger = (\rho, i\mathbf{j}) \cdot (\Phi, -i\mathbf{A}) \tag{4.82}$$

and

$$AJ^\dagger = (\Phi, i\mathbf{A}) \cdot (\rho, -i\mathbf{j}) \tag{4.83}$$

sums to a scalar and real quantity:

$$\frac{1}{2}\left(JA^\dagger + AJ^\dagger\right) = (\rho\Phi - \mathbf{j}\mathbf{A}, 0). \tag{4.84}$$

Based on all this the variational principle of electrodynamics as a functional of the quaternion fields and currents, in Lorenz gauge, takes the form

$$S = \int d^4x \, \frac{1}{2}\left(F^\dagger F + JA^\dagger + AJ^\dagger\right). \tag{4.85}$$

This form is simple and elegant. Moreover the wave equations can be derived from this easily. Only the basic properties of the quaternion multiplication and conjugation have to be used for this. Since the conjugate of a product is the product of the conjugates in opposite order (cf. the simplest quaternion representation contains matrices), the conjugate quaternion field strength can be expressed by using the derivation operator "acting backwards":

$$F^\dagger = \left(\partial^\dagger A^\dagger\right)^\dagger = A \overset{\leftarrow}{\partial} . \tag{4.86}$$

Here the partial derivation as a quaternion stands after the four potential quaternion, but it acts on the components of A, as "backwards". That is denoted by the backward arrow over the partial derivative operation quaternion.

Using all this we obtain for the part independent of sources

$$S_0 = \frac{1}{2} \int d^4x \; F^\dagger F = \frac{1}{2} \int d^4x \; \left[(A \overleftarrow{\partial}) (\partial^\dagger A^\dagger) \right]. \tag{4.87}$$

Integrations by part in this notation can be arranged by a single step. The "backwards" derivative becomes then a "forward" derivative:

$$S_0 = \frac{1}{2} \int d^4x \; \left[A(\partial \, \partial^\dagger) A^\dagger \right]. \tag{4.88}$$

In this form only the square length of the partial derivative operation quaternion occurs, this is a scalar operator.

$$\Box = \partial \, \partial^\dagger = (\partial_t^2 - \nabla^2, \, 0) \tag{4.89}$$

This is named to D'Alambert or sometimes to wave operator. The quaternionic action appears now in the following form:

$$S = \frac{1}{2} \int d^4x \; \left[-A\Box A^\dagger + AJ^\dagger + JA^\dagger \right], \tag{4.90}$$

This functional finally can be varied by A and $A^\dagger$ independently to obtain

$$\frac{\delta S}{\delta A} = -\Box A^\dagger + J^\dagger = 0,$$
$$\frac{\delta S}{\delta A^\dagger} = -\Box A + J = 0. \tag{4.91}$$

The result of variation against the quaternion four vector potential is a pure wave equation for all components (since $\Box$ is a quaternion scalar operation): $\Box A = J$.

Viewing the electromagnetic field strength as a quaternion plays a useful role beyond the elegant formulation for the action and the simple derivation of the wave equation. It can also be used for constructing the energy–momentum tensor components. If in the product with F we use the component-wise complex conjugated instead of the quaternion conjugate, $F^* - (G, i\mathbf{F}^*)$, then the energy density and current appears:

$$T = \frac{1}{2} FF^* = \frac{1}{2}(G, i\mathbf{F}) \cdot (G, i\mathbf{F}^*) \tag{4.92}$$

can be cast into the form

$$T = \frac{1}{2} \left(G^2 + \mathbf{F} \cdot \mathbf{F}^*, \; i G\mathbf{F}^* + iG\mathbf{F} - \mathbf{F} \times \mathbf{F}^* \right). \tag{4.93}$$

Here using the electric and magnetic fields in the complex field strength this expression simplifies to

$$T = \frac{1}{2} \left(G^2 + \mathbf{E}^2 + \mathbf{B}^2, \; 2iG\mathbf{E} + 2i\mathbf{E} \times \mathbf{B} \right). \tag{4.94}$$

In the Lorenz gauge $G = 0$, so what remain are only the familiar expressions of energy density and Poynting vector,

$$T = \left(\frac{1}{2} \left(\mathbf{E}^2 + \mathbf{B}^2 \right), \; i\mathbf{E} \times \mathbf{B} \right) = (w, i\mathbf{S}). \tag{4.95}$$

Chapter 5
Quantum Mechanics: The Most Classical Non-classical Theory

The fundamental law of quantum mechanics is summarized in the Schrödinger equation. In this chapter we derive this equation from a variational principle, as Schrödinger did it himself in his original publications. However, for the sake of playfulness we do not follow his paper, but we walk around the variational principle behind his equation. First the stationary case, then the time-dependent case, finally the case with relativistic energies will be investigated. The latter might shed some light onto why Schrödinger first tried to derive his famous equation from the Klein–Gordon equation.

We start with the Hamilton–Jacobi equation, that describes the classical, nonrelativistic dynamics:

$$H\left(Q, \frac{\partial S}{\partial Q}, t\right) + \frac{\partial S}{\partial t} = 0, \tag{5.1}$$

with S being the action, Q the generalized coordinate, t the time parameter and $H(Q, P, t)$ the Hamilton function.

5.1 Stationary Case

In the stationary case the Hamilton function does not depend on time explicitly, therefore the time dependence of the action is trivial and can be separated: $S = S'(Q, \dot{Q}) - tE$. The Q-derivative of the reduced action, S', is the same as that of the complete action. The Hamilton–Jacobi equation takes a simpler form,

$$H\left(Q, \frac{\partial S}{\partial Q}\right) - E = 0, \tag{5.2}$$

© The Author(s), under exclusive license to Springer Nature Switzerland AG 2023
T. S. Biró, *Variational Principles in Physics*,
SpringerBriefs in Physics,
https://doi.org/10.1007/978-3-031-27876-1_5

telling that the value of the Hamilton function is the energy. The classical motion in phase space happens on a hypersurface with constant energy (shell).

In the quantum mechanics a breaking of the classical dynamics is realized. The question is, how. Let us characterize the measure of the break with classical physics by the following integral,

$$K_{\text{stac}} = \int dQ \left[H\left(Q, \frac{\partial S}{\partial Q}\right) - E \right] w(Q), \tag{5.3}$$

and search for that theory, which minimizes this break. This minimization is a variational principle, that leads to the stationary Schrödinger equation, provided the assumptions below.

First we express the action in terms of the eikonal, ψ, as

$$S = k \ln \psi, \tag{5.4}$$

with k being a constant to be determined later. Although this formula reminds about the Boltzmannian definition of entropy, we should not mix it up. The action S is not an entropy, and the eikonal factor is not a probability or phase space volume. The weighting factor, $w(Q)$, in fact depends on the whole orbit; in simple cases only through the action. We search for that $w(\psi)$ weighting factor under the integral, which makes the variational minimum equation linear in ψ. This requirement of linearity corresponds to the experienced wavelike behavior, to the interference of possible orbits in quantum mechanics.

First we carry out this analysis for a certain class of Hamilton functions,

$$H(Q, P) = \frac{1}{2M} P^2 + V(Q). \tag{5.5}$$

This means a separable and nonrelativistic kinetic energy term. Our quantity to be varied is the following integral:

$$K_{\text{stac}} = \int dQ \left[\frac{k^2}{2M} \left(\frac{1}{\psi} \frac{\partial \psi}{\partial Q} \right)^2 + V(Q) - E \right] w(\psi). \tag{5.6}$$

In order to achieve a variational equation linear in ψ, the expression under the integral must be quadratic, so we choose

$$w(\psi) = \psi^2. \tag{5.7}$$

By doing so we obtain

$$K_{\text{stac}}^{\text{lin}} = \int dQ \left[\frac{k^2}{2M} \left(\frac{\partial \psi}{\partial Q} \right)^2 + (V(Q) - E)\psi^2 \right]. \tag{5.8}$$

The variational equation minimizing the above expression, quantifying the break with the classical dynamics, is nothing else than the stationary Schrödinger equation.

$$\frac{\delta K^{\text{lin}}_{\text{stac}}}{\delta \psi} = -\frac{k^2}{2M}\frac{\partial^2 \psi}{\partial Q^2} + (V(Q) - E)\psi = 0. \tag{5.9}$$

Solutions of this equation for $\psi = \exp(S/k)$ can be linearly combined.

Problems around the interpretation of quantum mechanics partially stem from insisting on the exact fulfillment of the Hamilton–Jacobi equation, in the stationary case simply meaning that the Hamilton function equals to the energy, even in the world of atoms. In the Copenhagen school and even beyond many interpret the Schrödinger equation so, that it were formally an energy equation,

$$\frac{1}{2M} P^2 + V(Q) - E = 0, \tag{5.10}$$

in general a Hamilton–Jacobi equation, only P carries a meaning and a role different from classical mechanics. This is mathematically equivalent to fixing a consequence of the variational principle of Schrödinger to a classical formula on the energy:

$$\left(\frac{1}{2M} \hat{P}^2 + V(\hat{Q}) - E\,\hat{1}\right)\psi = 0. \tag{5.11}$$

Here by the sign $\hat{\ }$ we emphasize that those quantities are no more functions or variables, but operators:

$$\hat{P} = \frac{k}{i}\frac{\partial}{\partial Q}, \qquad \hat{Q} = Q. \tag{5.12}$$

for consistency. A direct consequence is the uncertainty relation by Heisenberg:

$$\left[\hat{P}, \hat{Q}\right] = \frac{k}{i}. \tag{5.13}$$

Finally the value of the constant k is $\hbar$, the Planck constant divided by 2π.

5.2   Dynamical Case

In the dynamical case the variational principle contains a time integral, too. Wanting the linearity in ψ further, we have to allow for *complex* for it. A complex eikonal leads to a complex action, and therefore the momentum is also not necessarily real. This is the situation in the "prohibited" zone, where the kinetic energy is negative. This is in the classical physics forbidden.

Generalizing the discussion in the previous section, the weighting factor under the integral is modified to $w = |\psi|^2 = \psi^*\psi$. The term expressing the kinetic energy also changes,

$$T = \frac{1}{2M}\left|\frac{\partial S}{\partial Q}\right|^2 \geq 0, \tag{5.14}$$

is the non-negative expression. The break with the classical Hamilton–Jacobi equation is now integrated both over space and time:

$$K_{\text{din}} = \int dt \int dQ \left[H\left(\frac{\partial S}{\partial Q}, Q, t\right) + \frac{\partial S}{\partial t}\right] w(Q, t). \tag{5.15}$$

Owing to properties of Minkowski spacetime, here we seek an extremum only, not a minimum. Using the eikonal function ψ and multiplying with its complex conjugate we obtain

$$K_{\text{din}} = \int dt\, dQ \left[\frac{|k|^2}{2M}\frac{\partial \psi}{\partial Q}\frac{\partial \psi^*}{\partial Q} + V(Q)\psi\psi^* + k\psi^*\frac{\partial \psi}{\partial t}\right]. \tag{5.16}$$

Variation with respect to ψ^* results in

$$\frac{\delta K_{\text{din}}}{\delta \psi^*} = -\frac{|k|^2}{2M}\frac{\partial^2}{\partial Q^2}\psi + V(Q)\psi + k\frac{\partial \psi}{\partial t} = 0. \tag{5.17}$$

We can be satisfied with this result, if the complex conjugate of the variation with respect to ψ agrees with the variation result against the conjugate ψ^*:

$$\frac{\delta K_{\text{din}}}{\delta \psi} = -\frac{|k|^2}{2M}\frac{\partial^2}{\partial Q^2}\psi^* + V(Q)\psi^* - k\frac{\partial \psi^*}{\partial t} = 0. \tag{5.18}$$

After complex conjugation

$$\left(\frac{\delta K_{\text{din}}}{\delta \psi}\right)^* = -\frac{|k|^2}{2M}\frac{\partial^2}{\partial Q^2}\psi + V(Q)\psi - k^*\frac{\partial \psi}{\partial t} = 0 \tag{5.19}$$

which agrees with Eq. (5.17) only if

$$k = -k^*, \tag{5.20}$$

i.e. k is purely imaginary. The familiar time dependent Schrödinger equation emerges with the choice $k = \hbar/i$

$$i\hbar\frac{\partial \psi}{\partial t} = -\frac{\hbar^2}{2M}\frac{\partial^2}{\partial Q^2}\psi + V(Q)\psi. \tag{5.21}$$

Insisting formally on the classical dynamical equation, i.e. using operator versions $\hat{P}$ and $\hat{Q}$ one defines a Hamilton operator,

$$i\hbar\frac{\partial\psi}{\partial t} = H\left(\frac{\hbar}{i}\frac{\partial}{\partial Q}, Q\right)\psi = \hat{H}\psi. \tag{5.22}$$

We note that for a free motion $w = 1$ is constant and normalized all over the orbit.

5.3 Relativistic Case: Time-Independent Potential

The relativistic Hamilton–Jacobi equation for a mass point is based on the energy—momentum relation, the dispersion relation for the wave:

$$E^2 - (cP)^2 - (Mc^2)^2 = 0. \tag{5.23}$$

The presence of a time independent (static) potential modifies the energy, E to $E - V(Q)$. Ignoring a sign problem, related to the antimatter, now the energy is given by

$$E = \sqrt{(Mc^2)^2 + (cP)^2} + V, \tag{5.24}$$

and its nonrelativistic expansion up to order $1/c^2$ by

$$E = Mc^2 + \frac{1}{2M}P^2 + V. \tag{5.25}$$

This expression contains the usual kinetic and potential energy terms besides a rest mass energy, Mc^2. The corresponding Hamilton–Jacobi equation relies on the substitutions

$$P = \frac{\partial S}{\partial Q}, \qquad E = Mc^2 - \frac{\partial S}{\partial t}, \tag{5.26}$$

and obtains the form

$$\left(Mc^2 - V - \frac{\partial S}{\partial t}\right)^2 - \left(c\frac{\partial S}{\partial Q}\right)^2 - (Mc^2)^2 = 0. \tag{5.27}$$

Selecting terms according to Mc^2 factors the result is

$$\frac{\partial S}{\partial t} + \frac{1}{2M}\left(\frac{\partial S}{\partial Q}\right)^2 + V(Q) = \frac{1}{2Mc^2}\left(\frac{\partial S}{\partial t} + V\right)^2. \tag{5.28}$$

In the nonrelativistic approximation the right hand side (rhs) is zero. Physically this means that the kinetic energy

$$T = \frac{1}{2M} \left(\frac{\partial S}{\partial Q} \right)^2 \tag{5.29}$$

would be equal to $-\frac{\partial S}{\partial t} - V$, that—substituted into the rhs translates to the negligibility of the ratio kinetic to rest mass energy, T/Mc^2.

We allow for a complex eikonal in the relation $S = k \ln \psi$ learning from the experience of dynamical case. Therefore the factor, $w = |\psi|^2$, weights the break with the Hamilton–Jacobi equation, and the kinetic terms contain the absolute value squares of complex terms. The variational principle is to extremize the following functional:

$$K = \int dt\, dQ \left\{ \left| Mc^2 - V - \frac{k}{\psi} \frac{\partial \psi}{\partial t} \right|^2 - \left| c \frac{k}{\psi} \frac{\partial \psi}{\partial Q} \right|^2 - (Mc^2)^2 \right\} |\psi|^2. \tag{5.30}$$

Performing the multiplication and cancelling the $(Mc^2)^2$ terms we obtain

$$K = \int d^4x \left\{ -Mc^2 \left(2V\psi\psi^* + k\dot{\psi}\psi^* + k^*\psi\dot{\psi}^* \right) + |V\psi + k\dot{\psi}|^2 - |ck\nabla\psi|^2 \right\} \tag{5.31}$$

denoting the time-derivatives by overdot and the space derivation operator by the nabla symbol.

The functional derivative against ψ^* becomes

$$\frac{\delta K}{\delta \psi^*} = -Mc^2 \left(2V\psi + (k - k^*)\dot{\psi} \right) + \left(V^2\psi + (k - k^*)V\dot{\psi} - |k|^2\ddot{\psi} \right) + c^2|k|^2\nabla^2\psi. \tag{5.32}$$

Arranging the terms by their orders in c^2 we get

$$\left(-\frac{|k|^2}{2M}\nabla^2\psi + V\psi + \frac{k - k^*}{2}\dot{\psi} \right) = \frac{1}{2Mc^2} \left(V^2\psi + (k - k^*)V\dot{\psi} - |k|^2\ddot{\psi} \right). \tag{5.33}$$

In the nonrelativistic approximation the right hand side is zero, delivering the familiar Schrödinger equation. From the experience with quantum mechanics we distill the value $k = \hbar/i$, so the relativistic Schrödinger equation with a static potential, V, would have been the following:

$$\left(\hat{H} - i\hbar\frac{\partial}{\partial t} \right) \psi = \frac{1}{2Mc^2} \left(V - i\hbar\frac{\partial}{\partial t} \right)^2 \psi. \tag{5.34}$$

Here $\hat{H} = V + \hat{P}^2/2M$ is the usual nonrelativistic Hamilton operator. The operator of the kinetic energy, $\hat{T}$, the operator $\hat{A} = V - i\hbar\frac{\partial}{\partial t}$ and the notation $\epsilon = 1/2Mc^2$ will be now introduced, in order to arrive at a recursive formula:

$$\chi = \hat{A}\psi, \qquad \chi + \hat{T}\psi = \epsilon\hat{A}\chi. \tag{5.35}$$

If the operator $1 - \epsilon \hat{A}$ can be inverted and expanded into a convergent series, meaning the smallness of the kinetic energy compared to the rest mass energy, we obtain

$$\left(\hat{A} + \sum_{n=0}^{\infty} \epsilon^n \hat{A}^n \hat{T} \right) \psi = 0. \tag{5.36}$$

Another interesting limiting case occurs when $V = Mc^2$. With this choice Eq. (5.33) becomes the Klein–Gordon equation:

$$\frac{1}{c^2} \frac{\partial^2}{\partial t^2} \psi - \nabla^2 \psi + \left(\frac{Mc}{\hbar} \right)^2 \psi = 0. \tag{5.37}$$

It is also worth to discuss the extreme relativistic case, occurring for vanishing rest mass. Then from Eq. (5.32) we gain

$$\frac{1}{c^2} \ddot{\psi} - \nabla^2 \psi + \frac{2i}{\hbar c^2} V \dot{\psi} - \frac{1}{\hbar^2} V^2 \psi = 0. \tag{5.38}$$

Its solution is

$$\psi = \varphi e^{-\frac{i}{\hbar} t V}, \tag{5.39}$$

and for φ we obtain a wave equation describing propagation with the speed of light,

$$\frac{1}{c^2} \ddot{\varphi} - \nabla^2 \varphi = 0. \tag{5.40}$$

This results holds also for time-dependent V potentials, with the only difference that in the exponent not $t V$, but $\int V dt$ appears.

5.4 Relativistic Case: Electromagnetic Potential

The above analysis is incomplete, it does not handle either the spin or the antimatter. The spin can be noticed in the presence of a magnetic field, the antiparticle carries an electric charge opposite to that of a particle. So the relativistic Schrödinger equation is even more interesting when a general electromagnetic vector potential is present.

In a relativistic dispersion relation both the energy and the momentum get shifted in external electromagnetic fields:

$$(E - e\Phi)^2 - (c\mathbf{P} - e\mathbf{A})^2 - (Mc^2)^2 = 0. \tag{5.41}$$

The variational principle changes to

$$K = \int d^4x \left\{ \left| Mc^2\psi - e\Phi\psi - k\dot\psi \right|^2 - |ck\nabla\psi - e\mathbf{A}\psi|^2 - \left| Mc^2\psi \right|^2 \right\} \quad (5.42)$$

Variation against ψ^* is now equivalent with the complex conjugate of the variation against ψ for arbitrary complex k. It provides

$$\frac{\delta K}{\delta\psi^*} = \left\{ (Mc^2 - e\Phi)^2\psi - (Mc^2 - e\Phi)k\dot\psi + k^*\frac{\partial}{\partial t}\left((Mc^2 - e\Phi)\psi \right) - |k|^2\ddot\psi \right\}$$
$$- \left\{ -c^2|k|^2\nabla^2\psi - cke\mathbf{A}\nabla\psi + ck^*\nabla\left(e\mathbf{A}\psi \right) + e^2\mathbf{A}^2\psi \right\}$$
$$- \left\{ (Mc^2)^2\psi \right\} \quad (5.43)$$

Using the Lorenz gauge fixing, $c\nabla\mathbf{A} + \dot\Phi = 0$, so derivatives of the scalar and vector potential can cancel each other. The rest can be comprised into the following equation,

$$c^2|k|^2\nabla^2\psi - |k|^2\ddot\psi + (k - k^*)\left((Mc^2 - e\Phi)\dot\psi - ec\mathbf{A}\nabla\psi \right)$$
$$+ \left(-2Mc^2 e\Phi + e^2(\Phi^2 - \mathbf{A}^2) \right)\psi = 0. \quad (5.44)$$

With the choice $k = \hbar/i$ this result in

$$i\hbar\dot\psi = \left(-\frac{\hbar^2\nabla^2}{2M} + e\Phi \right)\psi + \frac{i\hbar e}{Mc}\mathbf{A}\nabla\psi + \frac{1}{2Mc^2}\left(\hbar^2\ddot\psi + e^2(\mathbf{A}^2 - \Phi^2)\psi \right). \quad (5.45)$$

This form, not taking into account terms with $\ddot\psi$ and Φ^2, coincides with the Pauli equation used with the gauge fixing $\nabla\mathbf{A} = 0$:

$$i\hbar\dot\psi = -\frac{1}{2M}\left(\frac{\hbar}{i}\nabla - \frac{e}{c}\mathbf{A} \right)^2\psi + e\Phi\psi + \frac{1}{2Mc^2}\left(\hbar^2\ddot\psi - e^2\Phi^2\psi \right). \quad (5.46)$$

It is rewarding to rewrite Eq. (5.45) using the $g = e/\hbar c$ coupling constant and the $\mu = Mc/\hbar$ inverse Compton wavelength parameters. It is

$$\frac{1}{c^2}\ddot\psi - \nabla^2\psi + 2g\Phi\,\mu\psi = 2i(\mu\frac{1}{c}\dot\psi - g\mathbf{A}\nabla\psi) + g^2(\Phi^2 - \mathbf{A}^2)\psi. \quad (5.47)$$

The massless limit appears at $\mu = 0$:

$$\frac{1}{c^2}\ddot\psi - \nabla^2\psi = -2ig\mathbf{A}\nabla\psi + g^2(\Phi^2 - \mathbf{A}^2)\psi. \quad (5.48)$$

The zero coupling limit looks like

$$\frac{1}{c^2}\ddot\psi - \nabla^2\psi = 2i\mu\frac{1}{c}\dot\psi, \quad (5.49)$$

or rearranged somewhat,

$$i\hbar\dot{\psi} = \frac{\hbar^2}{2M}\partial^2\psi, \tag{5.50}$$

with

$$\partial^2 = \frac{1}{c^2}\frac{\partial^2}{\partial t^2} - \nabla^2 \tag{5.51}$$

being the D'Alambert wave operator. In the limit $c^2 \to \infty$ we gain back the free Schrödinger equation. At the same time the solution of the wave equation (5.50),

$$\psi = \varphi\, e^{\frac{i}{\hbar}Mc^2t}, \tag{5.52}$$

with φ satisfying the Klein–Gordon equation

$$\frac{1}{c^2}\ddot{\varphi} - \nabla^2\varphi + \left(\frac{Mc}{\hbar}\right)^2 \varphi = 0. \tag{5.53}$$

5.5 Speculations

From the variational principle approach the Copenhagen interpretation does not follow. Here $w = |\psi|^2$ is a weighting factor in measuring the break with the classical dynamics, not a probability density. Therefore it is not guaranteed that its integral over the space would be finite, so whether the wave function is "normalizable" is a further question. On the other hand, theoretically, it may differ from zero on disjoint intervals. In that way regions may exist in spacetime where $w = 0$ and therefore arbitrary deviations from the classical Hamilton–Jacobi equation are allowed.

From the classical definition of the eikonal it follows that if S/k is purely imaginary, i.e. if using $k = \hbar/i$ the action, S, as is real then the weighting factor is unity: $w = 1$. This means that all trajectories with this property count with the same factor with respect to the variational principle. To the contrary, in typical classically forbidden zones, e.g. with negative kinetic energy, the momentum is purely imaginary and so is the $S = \int P dQ$ action. For such regions the weighting factor is less than one, and it decreases exponentially with the intrusion depth. Here the break with the classical dynamics is less important, so can be bigger. This is the physics behind quantum tunneling.

The following is speculative, but possibly a challenging exercise for the fantasy. Let us investigate step by step the thoughts summarized above by looking for mathematical alternatives and their physical interpretation. One question is why the constant k cannot be real in connecting the action and the eikonal. Or genuinely complex instead of pure imaginary. Perhaps the real part is suppressed by a c^2 factor. We emphasize that for a real $k = b$ the Schrödinger equation would not emerge, since the term describing imaginary diffusion would be missing. On the other hand it would for sure be causal, since the term describing the wave propagation in the relativistic treatment contains $|k|^2$ only. Then the functional derivative of the variational principle (5.32) would deliver

$$\frac{1}{c^2}\ddot{\psi} - \nabla^2\psi + \frac{1}{b^2}V(2Mc^2 - V)\psi = 0. \tag{5.54}$$

This equation gives wavelike solution in the range $V < 2Mc^2$, and exponentially damped ones for $V > 2Mc^2$. $V = 2Mc^2$ is the threshold for pair creation, above this the Klein paradox occurs . Here quantum field theory starts improving on quantum mechanics.

Speculations about what more can be done based on a variational principle are also possible. Formulating the Hamilton–Jacobi equation in gravitational field the above recipe may be applied to seek for an equation breaking minimally the classical theory in a variational sense. It is also an interesting question whether the Einstein–Hilbert action possesses a wave like limit, describing perhaps a post-newtonian gravity wave, where the above strategy could be repeated. This would be a Schrödinger variational type approach to quantum gravity. It is not clear on this level whether this could be related to string theory or not.

We formulate it again: the breaking of the equation describing the classical dynamics with a weight factor in spacetime so that the variational problems becomes linear in something (we have discussed so far the complex eikonal), truly contains quantum uncertainty. Obviously the Schrödinger equation, solving the variational problem of breaking with the Hamilton–Jacobi equation, is not identical with the Hamilton–Jacobi equation itself. When and if someone insists on the validity of the Hamilton–Jacobi equation (HJ), then that person would be at pains to transform the Schrödinger equation to a HJ form. The price to pay is the operator formalism. And so happened in the history of physics in the present universe.

The rest follows shortly. The operators $\hat{P}$ and $\hat{Q}$ cannot be exchanged, and therefore due to an algebraic identity their variances are correlated: the product cannot be made smaller than $\hbar/2$ at each component. Consequently the operator $\hat{H}$ is not exactly the physical energy, it "quantum fluctuates". The variational principle, we were discussing in this chapter, also says $H \neq E$, smeared in space but as a total minimally.

This view might help to resolve a contradiction between point particles and plane waves. The relativistic improvements on Schrödinger equation on the other hand ensure that no physical effect propagates faster than light, so no EPR paradox arises.

Further Readings

- K. Simonyi, *A fizika kultúrtörténete* (Gondolat Kiadó, Budapest, 1978).
  K. Simonyi, *A Cultural History of Physics*, translated from Hungarian to English by David Kramer (CRC Press Taylor & Francis Group, Boca Raton, FL, USA, 2012).

- C. Lanczos, *The Variational Principles of Mechanics* (University of Toronto Press, Toronto, 1949).

- I.M. Gelfand, S.V. Fomin, *Calculus of Variations* (translated to English by R.A. Silvermann) (Prentice-Hall Inc., Englewood Clitts, N.J., 1963) (Reprinted by Dover Publication, New York, 2000).

- L.D. Landau, E.M. Lifschitz, *Course on Theoretical Physics*. Mechanics, vol. 1, 3rd edn. (Butterworth-Heinemann, 1976).

- J.D. Jackson, *Classical Electrodynamics*, 3rd edn. (Wiley, 1998). https://www.academia.edu/42972797/Jackson_-_Classical_Electrodynamics_3rd_edition

- E. Schrödinger, Quantisierung als eigenwertproblem. Annalen der Physik (Leipzig) **361**(4), 361 (1926).

- R.P. Feynman, R. Leighton, M. Sands, *The Feynman Lectures on Physics*. https://feynmanlectures.caltech.edu/I_toc.html, https://feynmanlectures.caltech.edu/II_toc.html, https://feynmanlectures.caltech.edu/III_toc.html

- J.T. Deversee, G.V. Berghe, *Magic is No Magic; The Wonderful World of Simon Stevin* (WIT Press, Antwerpen, 2008).

© The Author(s), under exclusive license to Springer Nature Switzerland AG 2023
T. S. Biró, *Variational Principles in Physics*,
SpringerBriefs in Physics,
https://doi.org/10.1007/978-3-031-27876-1

- *Kleinsches Paradoxon*, Lexikon der Physik 3, Ha-Mh, 245 (Spektrum Akademis-cher Verlag, Heidelberg, 1999).

- D. Bleecher, *Gauge Theory and Variational Principles* (Addison-Wesley Pub. Co., Reading MA, 1981) (Reprinted by Dover Publication, New York, 2005).

Glossary of People Appearing in the Hungarian Book

A

Aristotle 384–322 BC, Greek philosopher, apprentice to Plato, private teacher of Alexander the Great. His most known works: Categories, Hermeneutics, First Analytics, Second Analytics, Topics, Physics, Meteorology, Zoology, Rhetorics, Etics, Poetics.

D'Alambert, Jean le Rond 1717–1783, French physicist, mathematician, philosopher. Took part in the preparation of the French Encyclopedia, materialist. Since 1754 member of the French Academy, starting in 1772 permanent secretary. The D'Alambert principle in its form known to date was written up by Lagrange.

Ampére, André Marie 1775–1836, French physicist and mathematician, one of the founding fathers of the theory of electricity, the unit of electric current strength is named after him. The Ampére-law is also associated to his name, describing the magnetic effect of electric currents.

B

Bernoulli family of Swiss mathematicians in the 18th century. *Daniel Bernoulli* (1700–1782) constructed the Bernoulli principle, *Jakob Bernoulli* (1654–1705) gave name to the Bernoulli numbers. Also known *Johann Bernoulli* (1667–1748), *Johann III Bernoulli* (1744–1807), *Nicolas I Bernoulli* (1687–1759) and *Nicolas II Bernoulli* (1695–1726). The Bernoulli equation, describing the currents in fluids, was published by Daniel Bernoulli in 1735.

© The Author(s), under exclusive license to Springer Nature Switzerland AG 2023
T. S. Biró, *Variational Principles in Physics*,
SpringerBriefs in Physics,
https://doi.org/10.1007/978-3-031-27876-1

Blanusa, Danilo 1903–1987, professor of mathematics in Zagreb, Croatia. Among others, he dealt with the four colour conjecture, in physics he is known from disputes over the temperature and energy of bodies moving at relativistic speeds.

Boltzmann, Ludwig Eduard 1844–1906, professor of physics in Vienna, world famous personality in thermodynamics, creator of the kinetic theory of atomic gases and the classical entropy formula. The Boltzmann equation describes the time evolution of the momentum distribution of colliding particles by using the "Stosszahlansatz" (a postulate about the number of collision). His name occurs in the Boltzmann constant, in the Boltzmann (or Maxwell-Boltzmann, sometimes Boltzmann-Gibbs) distribution and in the Stefan-Boltzmann coefficient.

Bohr, Niels 1885–1962, Danish physicist, one of the pioneers of quantum physics. The Bohr model was the first non-classical model about atoms. His name appears in phrases like Bohr radius, and Bohr magneton, too. Infamous are his disputations with Einstein about the essence and interpretation of quantum uncertainty. The element 107 is named Bohrium. A crater on the Moon and the asteroid 3948 carry his name. The Niels Bohr Institute in Copenhagen is renown among physics research institutes.

Born, Max 1882–1970, German physicist, important participant in the development of quantum mechanics. His work in solid state physics and optics is also important, he was awarded by the Nobel Prize in 1954.

Brown, Robert 1773–1858, a Scottish botanist, he discovered the random motion of pollen in a water drop under the microscope in 1827. The theoretical description of the Brownian motion was given by Albert Einstein in 1905.

Buridan, Jean 1295–1358, by his Latin name Johannes Buridanus, French vicar, who helped the distribution of the Copernican theory in Europe. As the most important philosopher in the late Medieval, he founded the notion of inertia. The theory of impetus is assigned to him, and—more known—a parabola about a hungry asinine not being able to choose the shortest path to the food.

C

Christoffel, Elvin Bruno 1829–1900, German physicist and mathematician. His name is carried fort by the Christoffel symbols and by the Riemann–Christoffel tensor in differential geometry.

Clausius, Rudolf Julius Emanuel 1822–1888, German physicist. His achievements are unforgettable in his works in classical thermodynamics, in particular he introduced the phrase "entropy" into physics.

D

Dirac, Paul Adrian Maurice 1902–1984, British physicist, one of the founders of quantum mechanics, brother-in-law of Eugene Wigner. He received the Nobel Prize in 1933 together with Schrödinger, "for the discovery of new effective forms of the atomic theory". It binds to his name the Dirac equation, which predicted the antiparticle to the electron, the positron, first. The Dirac distribution, also called Dirac delta, is frequently used in mathematical physics. The statistics of fermions is often cited as the Fermi–Dirac distribution.

Descartes, René 1596–1650, French philosopher, mathematician, artillery officer, founder of the Cartesian way of thinking. His most renown phrase, "cogito, ergo sum" (I think therefore I am) summarizes the basic principle of rationalism. He also founded the analytic geometry, the Descartes (or Cartesian) coordinate system enables us to describe curves by algebraic equations.

Doppler, Christian Andreas 1803–1854, Austrian mathematician and physicist. He taught in Prague, in the Hungarian Royal Mining Academy to Selmec, and finally in Vienna. He is most known due to the Doppler effect; most speed measuring devices, either radar or ultrasound based, apply his formula.

E

Einstein, Albert 1879–1955, Germany born, then Swiss, later Prussian physicist; he escapes from Nazi Germany to the USA. He is most known due to his theories of special and general relativity, however, he was awarded the Nobel Prize in 1905 for explaining the photoeffect. He was first in explaining the Brown motion, and pioneered the photon theory of light. His name is composed into the Bose–Einstein distribution and into the element Einsteinium. A pacifist, pantheist, worked long on the final unified field theory. His disputes with Bohr about the interpretation of quantum mechanics (according to Einstein God does not throw dice), as well as his role in promoting the construction of the first atomic bomb in his letter to president Roosevelt contributed to his world fame. The position of being the first president to modern Israel was offered to him, but he refused.

Eötvös, Loránd (Roland) 1848–1947, Baron, Hungarian physicist. Most known are his experiments with the Eötvös pendulum for measuring the strength and gradients of the gravitational acceleration, but he worked on the capillarity of fluids, too. He was Chairman of the Hungarian Academy of Science and minister for education. His name is carried by the Eötvös Physical Society, and the Roland Eötvös University in Budapest. The mineral lorandit is named after him. The CGS unit for the gravity gradient is 1 eotvos.

Euler, Leonhard Paul 1707–1783, mathematician and physicist with Swiss origin, but he spent the majority of his life on German ground and in Russia, Saint Petersburg. His results are important for the fields of mathematical analysis, number theory, geometry and graph theory. In physics, the Euler equation describes the streaming of fluids, while the Euler–Lagrange equations are the derived equations of motion to a Lagrangian.

F

Faraday, Michael 1791–1867, English physicist and chemist. He discovered the electromagnetic induction and created the first electric motor. He also investigated the chemical effect of the electric current. The SI unit for capacity, the farad, is named after him. A renown saying of him is that "One day Sir, you may tax it". It was his answer to William Gladstone, that time finance minister of Britain, when he asked about the profit from electricity.

Fermat, Pierre de 1601–1665, French mathematician. Most known due to the Fermat conjecture, which could have been proven only 300 years later, in the recent past. His interest included the calculation of probability and the prime numbers. In physical optics the principle of shortest light ray propagation time is named as Fermat principle.

Feynman, Richard Phillips 1918–1988, American physicist. In his young years he took part in the Manhattan project, in 1965 won the Nobel Prize (together with Schwinger and Tomonaga) for developing quantum electrodynamics. Several theoretical tools carry his name in physics: most known are the Feynman graphs, the Feynman path integral, the Feynman–Kac formula, the Hellmann–Feynman theorem, the Feynman parameterization of integrals in particle physics.

Fock, Vladimir Alexandrovich 1898–1974, Soviet physicist. Most known from the many body quantum mechanics; his name is connected to the phrase "Fock space" (and to corresponding operators), the Fock representation and the Hartree–Fock method.

Fokker, Adriaan 1887–1972, physicist and musician in Netherlands. He was born in Holland East India (to date Indonesia), he is a cousin of the airplane constructor Anthony Fokker. In physics he is most known due the Fokker–Planck equation, but he was involved in research about the geodesic precession in the theory of general relativity, too.

Fourier, Jean Baptiste Joseph 1768–1830, French physicist and mathematician. His most renown creation is the Fourier transformation, that he worked out for solving the heat diffusion problem, comprised in the Fourier equation. (He is not related to the utopist socialist, Francois Marie Charles Fourier.)

G

Galerkin, Borisz Grigorjevics 1871–1945, Belorus mathematician and engineer. Most known due to the Ritz–Galerkin method, often used in modern finite element algorithms.

Galilei, Galileo 1564–1642, the most renown Italian physicist. The foundation of mechanics, the study of free fall, the relativity principle for motion is tagged to his name as well as the discovery of the four largest moons around the Jupiter with a self-assembled telescope. He discovered the phases of Venus and the sunspots, he was a resolute promoter of the Copernican worldview. The Holy Inquisition made him to withdraw his theses, admittedly after this he commented that "eppur si muove" (yet it moves, i.e. the Earth). The CGS unit of the gravitational acceleration is 1 gal, to honor him.

Gauss, Johann Carl Friedrich 1777–1855, German mathematician and physicist, the Princeps Mathematicae. Most known from the Gauss theorem, but he dealt with the complex analysis, differential geometry, number theory, optics and magnetism, too. His name occurs in the Gauss (the bell shape) distribution, the Gauss method of least squares, the Gauss elimination by solving large systems of linear equations, moreover the unit of the magnetic field strength is also 1 Gauss. According to an anecdote, he was 6 when punished in elementary school by adding the numbers from 1 to 100. He solved this task in a minute, describing the numbers in the reverse order below, noting that all sums are 101, and there are 100 of them, and finally he divided by two to get the final correct result.

Gibbs, Josiah Willard 1839–1903, American physicist, chemist and mathematician. His name is associated with the foundation of modern thermodynamics, among others in the Gibbs distribution. He pioneered the use of vector analysis in theoretical physics and chemistry. It is to be mentioned the Gibbs–Duhem relation and the Gibbs free energy.

Gyarmati, István 1929–2002, Hungarian physicist, former member of Hungarian Academy for Science. Gyarmati established a variational principle for the thermodynamics of irreversible processes to describe the space-time evolution of dissipative transport phenomena. From his variational principle the linear transport equations can be derived: Fourier equation for heat conduction, Fick equation for diffusion, Navier–Stokes and Reynolds equations for streaming. His work synthesized the results of Onsager and Prigogine in general variational principle.

H

Hamilton, William Rowan, Sir 1805–1865, Irish physicist, astronomer and mathematician. In physics his name is most known due to Hamiltonian mechanics, but a central quantity in quantum mechanics, the Hamilton operator, also carries his name. He initiated the mechanical action principle, the Hamilton–Jacobi equation, the use of the phrase "tensor", the application of the "nabla" symbol, the quaternion algebra and some further theorems in algebra and group theory, like the Cayley–Hamilton theorem, the Hamiltonian path in diagrams and more.

Hartree, Douglas Rayner 1897–1958, English physicist and mathematician. In physics mostly known due to the Hartree–Fock method.

Heisenberg, Werner Karl 1901–1976, German physicist. He belongs to the founding fathers of quantum mechanics, in his young years he was an acolyte of Arnold Sommerfeld, Max Born and Niels Bohr. The fundamental uncertainty relation is assigned to him, as well as the commutator relations in its background. He made important contribution to the theory of ferromagnets by the Heisenberg model. He introduced the notion of isospin.

Helmholtz, Hermann Ludwig Ferdinand von 1821–1894, German physicist and medical doctor. He worked on the color vision and depth perception, on the conservation of energy, on electrodynamics and thermodynamics. His name is preserved in the Helmholtz–equation, in Helmholtz free energy, in Helmholtz spool and in the Helmholtz theorems. His hydrodynamical theorems describe the behavior of vortices in weakly viscous fluids.

Hilbert, David 1862–1943, German mathematician. His name lives on in Hilbert space, a fundamental structure for quantum mechanics. He is also renown due to the list about the ten most difficult mathematics problems in 1900, that he later extended to a list of 23. Among them the 3.a from the 10 (6. from the 23), called the "axiomatisation of physics", and some other are unsolved until to date.

I, J

Jacobi, Carl Gustav Jacob 1804–1851, Prussian mathematician. Physicist know his name by using the determinant Jacobian. Still, the Jacobi identity, the Jacobi bracket and the Jacobi elliptic functions together with the corresponding integral are memorable as well as the Hamilton-Jacobi equations.

K

Kepler, Johannes 1571–1630, German astronomer and mathematician, the Astrologist to the Royal Court of Rudolph II, emperor of the Holy Roman Empire. Kepler's laws comprise the motion of planets on elliptic orbits into quantitative rules.

Klein, Oskar Benjamin 1894–1974, Swedish physicist, professor in Stockholm starting 1930. He dealt with the theory of general relativity, cosmology and particle physics. His name is attached to the Klein–Gordon equation describing the waves of a free, relativistic scalar field. The Klein paradox, paraphrased according to the hole theory of Dirac, also wears his name.

L

Lagrange, Joseph-Louis 1736–1813, Italian-born French physicist, mathematician, astronomer. In physics he is known mainly due to Lagrangian, but also the Euler–Lagrange equation, the Lagrange multipliers and the Lagrange points in space orbits wear his name. His renown in number theory and mathematical analysis, too.

Langevin, Paul 1872–1946, French physicist. Most known is the Langevin equation, developed by him to describe the stochastic Brownian motion. He dealt with para- and diamagnetism, and also took part in the development of submarine detection by ultrasound. The twin paradox in special relativity stems from him.

Legendre, Adrien-Marie 1752–1833, French mathematician. In physics he is known due to the Legendre transformation, however he achieved a number of results also in algebra, number theory, statistics and analysis. Legendre polynomials and a crater on the Moon carry his name.

Levi-Civita, Tullio 1873–1941, Italian mathematician, most known are his works in tensor calculus. The totally antisymmetric unit tensor wears his name.

Lorentz, Heindrik Antoon 1853–1928, Nobel Prized physicist from Holland. His name is remembered in physics by the Lorentz transformation, and by the Lorentz force acting on charges which move in electromagnetic fields. He received the Nobel Prize in 1902 for the theoretical explanation of the Zeeman effect.

Lorenz, Ludvig 1829–1891, Danish physicist and mathematician. In optics the Wiedemann–Franz–Lorenz law, in physical field theory the Lorenz gauge wears his name. The relation between the rarefaction index and the density of the medium was discovered by him and by Hendrik Lorentz independently. Nowadays it is referred to as the "Lorentz–Lorenz equation".

M

Maupertuis, Pierre-Louis Moreau de 1698–1759, French mathematician and philosopher. He was the first chairman of the Scientific Academy in Berlin. His name is preserved in the Maupertuis principle determining mechanical orbits by a variational action.

Maxwell, James Clerk 1831–1879, Scotch theoretical physicist and mathematician. He was the creator of the classical field theory of electromagnetism. His name is carried fort in the Maxwell equations, in the Maxwell–Boltzmann distribution of atomic velocities in a gas, and the Maxwell demon: a fictional being, able to gain information about individual atoms and—based on that—to separate the slow atoms from the cold ones. In this way the second law of thermodynamics is supposed to be invalidated. However, as it was shown later by Leo Szilard, this being becomes hot, since by gathering information his entropy increases.

N

Navier, Claude-Louis 1785–1836, French physicist and engineer. His name is preserved in the Navier–Stokes equation. Since 1824 member of the French Academy, he invested ample time in researching elasticity.

Newton, Isaac, Sir 1643–1727, the most known English physicist of all times, moreover he was an astronomer, mathematician and philosopher, as well as an alchemist and theologist. His central work, "Philosophiae Naturalis Principia Mathematica" (mathematical principles of natural philosophy) lays down the basic principles of Newtonian mechanics. Besides the Newton equation further axioms laid by him in the description of mechanical motion and his law of general mass gravitation unified the earthly and celestial physics. Known are his experiments in optics (prism, telescope), further his integral and differential calculus, "fluxion theory", at the fundaments of mathematical analysis, like the Newton–Leibniz theorem.

O

Onsager, Lars 1903–1976, Norvegian theoretical physicist and physical chemist. He won the Nobel Prize in 1968. In physics his name is conserved due to the Onsager relations, stating the cross correlation of thermodynamical forces and currents.

Oppenheimer, Robert J. 1904–1967, American physicist, the "father of atomic bomb", professional leader of the Manhattan project. Most known result from him is the Born–Oppenheimer approximation.

Ott, Heinrich 1894–1962, German physicist, student of Sommerfeld. His name is mostly known due to a discussion between Ott and Einstein, where they occupied diametric viewpoints about the relativistic temperature of moving bodies.

P

Pythagoras, of Samos 575–495 BC, ancient Greek ionic philosopher and mathematician. Most known due to the Pythagoras theorem, relating the side lengths of an orthogonal triangle. Based on his views about the cosmic harmony and the importance of mathematics in the classical antique a religious movement was founded. The proof that the square root of two is an irrational number and the phrase "harmony of spheres" are attributed to him.

Planck, Max 1858–1947, German physicist, born as Karl Ernst Ludwig Marx Planck, his name Marx he himself changed to Max later. Most famous is the Planck constant named after him. That quantity was introduced in the Planck law of black body radiation, and it had been revealed later that it is a fundamental constant in nature. Nobel Prize winner (1918), most known promoter of Albert Einstein. The German research institution network, formerly Kaiser Wilhelm Institutes, are named to date to Max Planck Institutes. His name is worn in the Planck length, also as Planck scale, Planck mass, Planck time, Planck energy, Planck temperature; beyond which the entanglement between gravity and quantum physics cannot be prevented any more.

Podolsky, Boris 1896–1966, Russian, later American physicist. Known from the EPR (Einstein–Podolsky–Rosen) paradox, originally formulated to demonstrate the incompatibility between quantum mechanics and special relativity. A closer investigation of the mechanism of the information propagation resolves this paradox.

Poynting, John Henry 1852–1914, English physicist. His name is known from the Poynting vector, assigned to the energy current flowing in electromagnetic fields.

Prigogine, Ilya Viscount 1917–2003, born in the Soviet Union, a Belgian chemical physicist with Russian origin. He won in 1977 the Nobel Prize for his results on dissipative structures. Later the theory of Self Organizing Criticality (SOC) was developed from those results.

R

Reynolds, Osborne 1842–1912, Irish physicist and engineer. He worked on fluid dynamics, his name is noted in the Reynolds number, the ratio of inertial and viscous forces, and in the Reynolds theorem.

Rényi, Alfréd 1921–1970, Hungarian mathematician. Perhaps most known is the entropy formula named after him, but he also achieved results to remember in combinatorics, number theory and in graph theory, e.g. the Erdős–Rényi model of random graphs. In Budapest a Mathematical Research Institute wears his name.

Riemann, Georg Friedrich Bernhard 1826–1866, German mathematician. In physics he is known due to general relativity theory, in mathematics due to his oeuvre in complex analysis and differential geometry. Among others the Riemann tensor, Riemann's zeta function, the Riemann–Stieltjes integral and the Cauchy–Riemann equations remind us to his achievements.

Ritz, Walter 1878–1909, Swiss theoretical physicist. He is known due to the Ritz method and the Rydberg–Ritz formula.

Rosen, Nathan 1909–1995, Israeli physicist. See the note at Podolsky.

S

Schrödinger, Erwin Rudolf Joseph Alexander 1887–1961, Austrian physicist, one of the founders of quantum mechanics. He was awarded by the Nobel Prize in 1933 for the Schrödinger equation named after him. One of his suggestions to demonstrate a paradox property of quantum mechanics when viewed in terms of classical physics became famous as "Schrödinger's cat", where the killing of a cat is decided by reading out the actual state of a single spin. Or in another version the realization of a radioactive decay.

Shannon, Claude Elwood 1916–2001, American electronic engineer and mathematician. His name is preserved in the formula for Shannon entropy, reflecting the coupling of thermodynamical entropy to the quantity of bitwise information. He won the Nobel Prize in 1939.

Stokes, George Gabriel, Sir 1809–1903, English physicist and mathematician. He was Lucas Professor of Physics at the University of Cambridge. First this position was filled by Newton, later also by Hawking. His name is most known from Stokes theorem and the Navier–Stokes equation, but he was occupied by investigating spectroscopy (Stokes and anti Stokes lines) and the fatigue of bridges.

T

Tsallis, Constantino, 1943–, Greek born, in Paris educated, Brazilian physicist. Most known due to the non-extensive entropy formula, promoted by him since 1988, as Tsallis entropy. This generalizes the classical entropy to systems showing anomalous fat tails in statistical distributions.

U, V

Uhlenbeck, George Eugene, 1900–1988, American physicist, born in Holland East India. In physics he is known by his modification to the Boltzmann equation, as

Uehling–Uhlenbeck blocking factors, as well as due to the stochastic Ornstein–Uhlenbeck process. He was a student of Ehrenfest, he received the Lorentz (1964) and Max Planck Medals (1970), and the Wolf Prize (1979).

Z

Zenon, 490–430 B.C., classical Greek philosopher before Socrates in Eleia, South Italy. Aristotle called him as the inventor of dialectics (the science of disputes). He is known due to his paradoxes about motion and time.

Printed in the United States
by Baker & Taylor Publisher Services